Cognitive and Neural Modelling for Visual Information Representation and Memorization

Cognitive and Neural Modelling for Visual Information Representation and Memorization

Limiao Deng

CRC Press

Taylor & Francis Group
Boca Raton   London   New York

CRC Press is an imprint of the
Taylor & Francis Group, an **informa** business

First edition published 2022
by CRC Press
6000 Broken Sound Parkway NW, Suite 300, Boca Raton, FL 33487-2742

and by CRC Press
4 Park Square, Milton Park, Abingdon, Oxon, OX14 4RN

CRC Press is an imprint of Taylor & Francis Group, LLC

© 2022 Limiao Deng

Library of Congress Cataloging-in-Publication Data
A catalog record has been requested for this book

ISBN: 978-1-032-24911-7 (hbk)
ISBN: 978-1-032-25119-6 (pbk)
ISBN: 978-1-003-28164-1 (ebk)

DOI: 10.1201/9781003281641

Typeset in Minion
by MPS Limited, Dehradun

Contents

## CHAPTER 1

Introduction

1.1 BACKGROUND

Humans can easily detect and recognize objects, no matter how different their appearance is and how complex the environment is; however, this basic function is a great challenge for computer vision systems. Humans can do this easily because they have a powerful system of visual perception and memory. Among the external information the brain receives, visual information accounts for a large proportion, about 80% of all perception information. The processing process in the visual system is very complex. Memory is very important for human beings. Because of memory, human beings have the ability to learn and remember what they have seen and experienced, which is conducive to rapid recognition of remembered scenes, cognition of new things, and adaptation to new environments. The main purpose of computer vision research is to enable computers to perceive, interpret, and understand the environment just as people do. However, the current level of computer vision is far from human vision. So, it is of important theoretical and applicational value to simulate the visual information processing of visual perception and brain memory mechanisms to achieve effective representation of visual information and rapid memory and to achieve high reliability, strong adaptability of image recognition and classification systems on the basis of the cognitive neuroscience research of visual perception and the brain memory mechanism[1].

DOI: 10.1201/9781003281641-1

How to process large amounts of visual information and extract useful information, and how to organize and represent the information effectively, have been a hot and difficult topic in the field of computer vision. The biological visual system shows excellent performance in this aspect. For example, in object recognition, no matter how the environment and light change, whether the object changes in scale, angle of view, rotation, position, and so on, the visual system can easily acquire the invariant characteristics of describing the object and perform rapid and effective recognition. Based on this, many researchers have tried to simulate the structure and function of visual cortex of the brain and constructed visual perception models based on biologically inspired methods to improve the processing ability of visual information. Although a great deal of achievements has been made in the modelling of biological visual perception, some of the visual perception models lack the description of biological characteristics, while some are too complex and fail to fully demonstrate the advantages of biological visual information processing[2].

Until now, the neurophysiological mechanism of human brain memory is still poorly understood, and the research on memory mainly comes from cognitive psychology. In order to simulate the process of memory in the human brain, researchers have carried out a large number of memory experiments and proposed many logical and computational memory models. However, at present, most memory models take vocabulary lists as research objects, and the characteristic expression of memory information is relatively simple, while the memory of visual information is rarely studied[3]. In recent years, on the basis of cognitive neuroscience research, some neural network models have been established to simulate how the cerebral cortex performs complex and necessary memory functions. However, most of these models are conceptual or abstract models, which are quite complicated to realize and cannot meet the actual processing requirements of natural images.

Therefore, on the basis of anatomical and cognitive neuroscience research, it is of great significance to study how visual information is represented, stored, and remembered in the brain and carry out cognitive neural modelling to realize reliable image classification and recognition. In this book, the visual pathway is modelled based on the mechanism of biological visual perception, and a method of constant feature extraction and expression of biological visual excitation is established. Then, based on the research results of anatomy and cognitive neuroscience, a visual memory model was established to simulate the process of the storage and extraction of visual information by the human brain, combining the memory mechanism of the brain and relevant theories of memory model.

The significance of this book is to break through the traditional visual information processing mode, combine the visual perception and memory mechanism of the human brain with computer vision and image processing technology, and provide a new way for image understanding, recognition and classification.

1.2 RESEARCH STATUS OF THE SUBJECT

1.2.1 Review of biological visual perception

The rapid development of brain science and cognitive technology and research methods makes it possible to study the neural mechanism of the structure and function of the biological visual system[1]. To explore the neural mechanism of the biological visual perception system and simulate the biological visual perception mechanism to improve the processing ability of computers, visual information processing has become a hot research topic at present.

The human visual system is a powerful system capable of distinguishing millions of different objects[4]. A comprehensive and accurate understanding of the structure and function of the visual system is necessary to build a computational model based on biological vision. Visual information is processed in a hierarchical way in the visual system and transmitted between cortical regions of the brain in accordance with a certain path. The path of visual information transmission is called visual pathway. Ungerleider et al.[1,5,6] believed that the visual system consists of two parallel visual pathways: the ventral pathway and the dorsal pathway, as shown in Fig. 1.1.

The ventral pathway starts from the retina, passes through the lateral geniculate nucleus (LGN), the visual cortex (V1, V2, V4) areas, the inferior temporal cortex (IT), and finally reaches the ventral prefrontal cortex (VLPFC). This pathway, also known as the What pathway, is responsible for processing static visual information and realizing object perception and recognition. The dorsal pathway starts from the retina, passes through LGN, visual cortex region (V1, V2), middle temporal region (MT), posterior parietal cortex (PPC), and finally reaches the

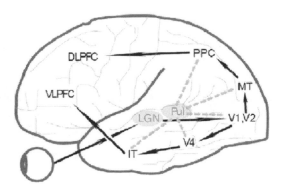

FIGURE 1.1   Two pathways of visual system.

dorsolateral prefrontal cortex (DLPFC). This pathway is also known as the Where pathway, which is responsible for processing dynamic visual information and realizing the recognition of spatial position and movement[7]. The pathway theory of the biological visual system is supported by a large number of anatomical and neurophysiological experimental data, which provides a theoretical basis and research basis for image recognition and classification research based on the mechanism of biological visual perception.

In the real world, there may be large differences between images of the same kind of objects, which are caused by the changes of shape and size within the class, as well as the interference of the shooting Angle, lighting conditions, and the environment. One of the main problems solved by biological vision system is to establish a representation of visual information, which can make object recognition free from interference of factors such as size, contrast, position, angle of view, and light. In order to ensure the accuracy of object recognition, the visual system must not be affected by intra-class differences and be sensitive to small differences between classes, that is, it has good intra-class invariability and inter-class specificity[1]. Therefore, how to simulate the powerful biological vision system to extract the invariant features of describing objects has become a key problem to be solved in image recognition and classification.

The ventral pathway is an important pathway for object recognition. The results of cognitive neuroscience research show that the ventral visual cortex usually organizes and processes information in a hierarchical manner, and these structural characteristics play a very important role in the establishment of biological visual perception model. In recent years, based on the relevant research achievements of neurophysiology and cognitive science, scholars at home and abroad have proposed many hierarchical computing models based on biological inspiration to simulate the basic principle and information processing process of the ventral visual pathway.

Neocognitron[8], proposed by Fukushima, was an early computational model of biological visual perception. Neocognitron is a hierarchical neural network model, whose structure consists of alternating simple neuron layer and complex neuron layer. Simple neuron performs convolution operation, while complex neuron is responsible for pooling and subsampling. The features extracted by this model have certain invariability and selectivity.

The more influential model of biological visual perception is the hierarchical maximization model, HMAX[9], which was first proposed by Riesenhuber and Poggio in 1999. The model is composed of a linear and nonlinear hierarchical structure, which is used to simulate the hierarchical processing mechanism of visual information in the ventral pathway and explain the information processing in the higher regions of the visual cortex of the brain. In 2007, Serre et al.[10] extended the HMAX model, added the learning process of feature template and template matching operation, and further refined the functions of each layer. Through the alternations of feature template matching and maximum pooling operation, visual representation of increasing complexity and invariability was obtained. Although the HMAX model has shown strong performance in object recognition, it has been successfully applied to real-time image classification[11,12], however, this model still has some limitations. For example, feature templates are randomly selected, and a large number of feature templates need to be extracted to ensure the performance of the model, which greatly increases the processing time. At the same time, although the model shows good scale and displacement invariance, it is sensitive to rotation[11-13]. At the same time, the model simulates the function of V1 neurons well but does not fully describe the biological characteristics of neurons in the upper regions (such as V2 and V4).

In the following years, many improvements have been made to improve the overall recognition performance of the HMAX model and to expand its application scope[14-24]. In 2008, Mutch and Lowe[14] improved the model by using several biofeasible methods, including the feature selection of sparse input, reaction inhibition, and final output. Meyes and Wolf[16] extracted a new feature set from the HMAX model for face processing and recognition. In 2011, Huang et al.[17] improved the HMAX model from two aspects, including eliminating useless input information under sparse constraint and using feature selection based on Boosting. To select an effective feature template, Ghodrati et al.[18] used genetic algorithm for template selection to improve the overall recognition performance of the model in 2012. Hormat et al.[19] introduced the maximum singular value into the HMAX model to improve the rotation invariance of features. In 2014, Lu et al.[20] adopted OGHM instead of Gabor filter to improve the rotation invariance of the model and improve the classification performance of the model to a certain extent. Tang and Qiao[21] extend the idea of maximizing operations to increase tolerance for differences within classes by proposing an overall framework with an HMAX model as the basic unit that can adaptively select the best matching template from a range of templates. In order to further improve the accuracy of model recognition, Alameer et al.[24] proposed an elastic network normalized dictionary learning method in 2015, so as to obtain relevant and informative features for object classification.

Another important model of biological visual perception is VisNet model proposed by Oxford University's Wallis and Rolls[25,26] in 2006. The model attempted to simulate the entire visual cortex of the brain involved in object recognition at the neuron level. The VisNet model consists of a series of competing neurons organized hierarchically, each layer containing short-term mutual suppression[27]. The model is trained in an unsupervised manner through improved association learning rules, which are combined with the time tracking of the activity of the front neurons, which are critical to the ability of neurons to have learning transformation invariance[28,29]. Based on the tracking rules, the competitive learning mechanism is established layer by layer, so as to realize the invariability of the changes of scale, position, and perspective.

In 2006, Hinton et al.[30] published a paper in Science, which started the upsurge of deep learning. He proposed Deep Belief Nets (DBN) and their corresponding efficient learning algorithms[31]. These algorithms have solved the difficulty of deep network training in the past and become the main framework of deep learning algorithm[32].

Convolutional Neural Networks (CNN)[33,34] are developed on the basis of traditional multi-layer perceptual Neural Networks. Its typical model LeNet-5 system[35] has an error rate of only 0.9% on MNIST data set. However, due to the lack of good effect on large images, it was once ignored. Until 2012, the deep CNN proposed by Krizhevsky et al.[36,37] achieved great success in ImageNet data set, making deep learning popular rapidly.

Deep neural networks exploit the property that many natural signals are compositional hierarchies, in which higher-level features are obtained by composing lower-level ones. In images, local combinations of edges form motifs, motifs assemble into parts, and parts form objects. Similar hierarchies exist in speech and text from sounds to phones, phonemes, syllables, words, and sentences. The pooling allows representations to vary very little when elements in the previous layer vary in position and appearance.

The convolutional and pooling layers in ConvNets are directly inspired by the classic notions of simple cells and complex cells in visual neuroscience[38], and the overall architecture is reminiscent of the LGN–V1–V2–V4–IT hierarchy in the visual cortex ventral pathway[39]. When ConvNet models and monkeys are shown the same picture, the activations of high-level units in the ConvNet explain half of the variance of random sets of 160 neurons in the monkey's inferotemporal cortex[40]. ConvNets have their roots in the neocognitron[41], the architecture of which was somewhat similar but did not have an end-to-end supervised-learning algorithm such as backpropagation. A primitive 1D ConvNet called a time-delay neural net was used for the recognition of phonemes and simple words[42,43].

There have been numerous applications of convolutional networks, such as image classification[44–46], face recognition[47,48], emotion recognition[49,50], Natural Language Processing (NLP)[51,52], etc.

Although the deep network has made great achievements in various fields, due to the complex structure of the deep network model, a large number of training data sets are needed in the process of model training, and a large number of free parameters need to be optimized through a large number of training, resulting in the excessively long training time and easy overfitting of the model. As a result, the popularization and application ability of the model is relatively poor, and there are still great challenges in engineering practice and application. How to improve the training algorithm of the model and greatly reduce the training time of the model is an urgent problem to be solved.

1.2.2 Review of brain memory model

With the continuous development of cognitive psychology and cognitive neuroscience, people's understanding and understanding of complex human brain activities have been gradually deepened, and a large number of new theories and methods have emerged in succession, which has promoted the continuous development of machine vision. To explore the neural mechanism of human brain memory and simulate its structure and function in order to improve the processing ability of computer vision has gradually become the focus of research at home and abroad[53]. The brain is an organ with information storage function, and its memory mechanism has received extensive attention from experts in cognitive psychology, cognitive neuroscience, and other fields[54,55].

Cognitive psychologists classify memories as transient, short-term, and long-term based on how long they last in the brain. Tulving[56] believed that memory was a system composed of independent modules, and long-term memory can be divided into declarative memory and non-declarative memory, while declarative memory can be further divided into episodic memory and semantic memory. Based on the research results of cognitive psychology, many abstract computational models have been proposed to describe the algorithms supporting recognition and memory.

In terms of episodic memory, Raaijmakers and Shiffrin[57] came up with the Search of Associative Memory model in 1981. In this model, information is divided into short-term store (STS) and long-term store (LTS), and the amount of information stored in LTS is related to the length of time the information is learned in STS. In the process of information extraction, the success of the stored information extraction depends on the degree of correlation between the extracted clues and the information. The probability of extracting certain information is the ratio of the intensity of association between the information and the extracted clue to the sum of the degree of association between the extracted clue and all stored information.

Based on the relevant theories of SAM model, Shiffrin and Steyvers[58] proposed REM model (normalization from Memory) in 1997. The main difference between The REM model and the previous models (such as

SAM model and MINERVA 2 model[59]) is that the REM model realizes bayesian computing of matching likelihood values between clues and stored memory tracks.

The REM model provides a mechanism for the memory system to respond to specific cues, but it does not explain how the memory system works without a detailed description of external cues[53]. To describe in detail how context changes over time and how context provides clues to memory during extraction, Howard and Kahana[60] proposed the Temporal Context Model in 2002, sorted memory contents according to time order, correlated memory with time factors, and explained decency effect and proximity effect.

In terms of semantic memory, the most famous one is Rumelhart model proposed by Rogers and McClelland[61]. The purpose of the Rumelhart model is to activate the correct set of attributes when probed with an item and association. One of the main features of the model is that it can be extended to new stimuli based on the similarity between the stimuli learned. In the fields of anatomy, physiology, and cognitive neuroscience, there are also a lot of research on human brain memory activities and mechanisms. Many years of research have proved that the Medial temporal lobe (MTL) is an important structure which plays a key role in memory. The research results from Burwell[62], Witter et al.[63] showed that the main MTL regions of humans, monkeys, and rodents have functional structures for memory processing. MTL region can be divided into the perinasal cortex, the parahippocampal gyrus, the entorhinal cortex, and the hippocampus. Most of the inputs to the perinatal cortex come from areas that process "What" information about object properties, while most of the inputs to the parahippocampal gyrus come from areas that process "Where" information about space. Subsequently, the "What" and "Where" information flows remain largely separate, with information flows from the perinasal cortex mainly going to the lateral entorhinal cortex, while information flows from the parahippocampal gyrus into the entorhinal cortex, Where the "What" and "Where" information flows converge. The output of hippocampal processing involves the return of feedback connections from the hippocampus to the entorhinal cortex, then through the perinasal and parahippocampal gyrus, and finally to the neocortex[64]. Based on this, the researchers propose a variety of memory models based on the anatomical physiological structure of the human brain.

Combining the results of neurophysiology and psychophysics research with neurocomputational methods, Rolls et al.[65,66] put forward the concepts of self-association and pattern association and gave a prototype system of self-association memory and pattern association memory network. In 2006, Rolls[67] proposed a computational theory of hippocampal function and predicted the function of different subregions of the hippocampus (dentate gyrus, CA1, and CA3), suggesting that autoassociative networks are essential for the production of episodic memory.

In 1995, McClelland et al.[68] proposed the theory of CLS of hippocampus structure and neocortex, believing that experience is generated by neocortex through slow Learning and that the hippocampus can quickly learn new projects. On this basis, Norman and O'Reilly proposed the CLS network model[69] in 2003, which respectively modelled the hippocampus and neocortex. It suggests that the neocortex is a distributed, overlapping system responsible for slow learning of semantic memory. The hippocampus is a sparse, patternless system responsible for the rapid learning of episodic memory. The model makes no assumptions about how the hippocampus and MTL regions, respectively, contribute to memory recognition. In 2010, Norman et al.[70] explained how hippocampus and neocortex contribute to memory recognition from the broad consensus of anatomical and physiological characteristics of hippocampus and neocortex.

Although CLS provides a neural computing framework that combines episodic and semantic memory, there are still some drawbacks. Bogacz and Brown[71] pointed out that the neocortex's ability to discriminate familiarity was much lower than that of human recognition of memory. Much of the problem was due to the fact that Hebbian's learning rules did not yet have a good enough sense of how to adjust synaptic weights. Moreover, Hebbian learning has a large capacity limit and therefore has less computing power than other error-based learning mechanisms[72]. At the same time, the CLS model is very complex, and the extension of this model makes the model more complex, and it makes little sense to explain some simple phenomena with an overly complex model[73].

In conjunction with a leading expert in cognitive neuroscience and computational intelligence, Sandia National Laboratories (SNL)[74] proposed a computational framework that addresses the cognitive neural mechanisms of how people remember past experiences. The SNL computational model inherits the concept of the important role of hippocampus in learning and memory, focuses on the important division of labour between hippocampus and cerebral cortex system described by CLS theory, and combines memory representation and processing. At the same time, it clearly illustrates how hippocampus and cerebral cortex represent each other in multi-level abstract concepts to support the cross storage of new information in cerebral cortex. This model is mainly realized by ART neural network[75]. The whole network is made up of many fuzzy ART modules, which have a complex structure.

In conclusion, although in the human brain memory mechanism modelling, the scholars in the field of cognitive psychology and cognitive neuroscience have made a lot of research results, but most of the proposed model to abstract calculation model, to implement more complex, and the object of study for the vocabulary list, the visual image of the learning and memory research is very few.

1.2.3 Review of Bayesian brain and free energy theory

Half a century ago, Ashby, a British psychiatrist and cybernetician, pointed out that the whole function of the brain boils down to error correction[76]. Since then, computational neuroscience has made great progress. Now there are more and more reasons to believe that Ashby is right. It is generally believed that error correction in mammalian brains is carried out through a series of cortical processing activities. By establishing causal systems in the objective world, higher-level systems attempt to predict the input of lower-level systems[77]. Such models have followed Helmholtz's[78] view that predictive perception is a probabilistic, knowledge-driven reasoning process. Helmholtz's views contributed to the influential work of Mackay[79], Neisser[80], and Gregory[81], namely the famous comprehensive analysis method.

Helmholz's views also played a very important role in the research work of computational science and neuroscience, and he made some important achievements in the field of machine learning, such as McClelland and Rumelhart's[82] pioneering work in back-propagation learning, and the Helmholz machine proposed by Dayan and Hinton[83,84] in 1995. Helmholz machines attempt to learn new expressions in a multi-layer system without providing a large number of pre-classified samples of input and output maps. It USES a top-down connection to provide the state of the hidden unit and USES a generation model that can create the perception pattern from me to oversee the generation of the perception recognition model. This strategy of using bottom-up connections to generate a virtual representation of the perceived data through a deep multi-layer network cascade is then used as the core of the hierarchical predictive coding approach to realize the perception[85-87].

In the 1990s, free energy formulas derived from physical statistics were introduced into the field of machine learning, and Hinton and Friston used free energy as a computable standard to measure the difference between the actual features of the real world and those captured by neural network models. In 1994, Hinton and Zemel[88] used the unbalanced Helmholtz free energy as the objective function for factor code learning in the automatic coding network. Then, MacKay[89] proposed a free energy minimization algorithm in 1995 and used it for decoding and cryptographic analysis. Neal and Hinton[90] proposed a function similar to negative free energy in 1998 and showed that in EM algorithm, the M stage maximizes the function from the perspective of model parameters, while the E stage maximizes the function from the perspective of distribution of variables, which has never been observed.

Friston et al. then combined the theory of free energy with the cognitive mechanism of the human brain and carried out a lot of research work. Friston and Stephan[91] believed that the perceptual process was only an emergent feature of the system following the principle of free energy. A system could change its Settings to minimize the free energy to achieve the purpose of influencing the environment or changing expectations. Meanwhile, Friston et al.[92] synthesized the relevant theories and believed that the Bayesian brain was derived from the general theory of free energy minimization. In this framework, behaviour and perception are considered to be a result of the suppression of free energy, leading to perceptual and behavioural reasoning that leads to a more specific Bayesian brain. Friston[93,94] believed that when the theory of free energy was applied to different elements of neural function, it could lead to the production of effective internal representation patterns and reveal the deeper basis behind the connection between perception, reasoning, memory, attention, and behaviour. Then, Friston and Kiebel[95] combined free energy theory and prediction coding behaviour to predict and explain the brain recognition, prediction, and perception classification for the brain as an inference problem, given a generation model of sensory input, can be based on the theory of freedom can call a generic method for model inversion, and shows that the brain has to implement the necessary mechanism of inversion.

Although Friston gave a framework for the free energy of cognitive and reasoning processes in the brain, there has been no concrete application of this framework so far.

1.3 MAIN CONTENT

In this book, focusing on the existing problems in the current research and referring to the relevant research results of cognitive neuroscience, biological visual perception models, and human brain memory models are established on the basis of the research in the mechanism of biological visual perception and the mechanism of human brain memory, so as to represent and memorize visual information and apply it to image classification.

The work of this book breaks through the traditional visual information processing method, combines the human brain perception and memory mechanism with computer vision technology, and tries to provide a new approach for image understanding, recognition and classification. Specifically, the chapter arrangement and main contents of this book are as follows:

Chapter 1 first introduces the research background and significance of this book and then summarizes the research status of biological visual perception, human brain memory modelling, and the theory of free energy of human brain cognition. Last, the main contents of this book are introduced.

Chapter 2 introduces the neural mechanism of biological visual perception and human brain memory, and the modelling methods of visual perception and memory in detail.

Chapter 3 proposed a bio-inspired model for object recognition based on Histogram of Oriented Gradients (HOG). HOG algorithm is integrated into the standard HMAX model. Meanwhile, a normalized dot-product operation is used to compute the S2 unit response instead of Euclidean distance. Furthermore, an effective template selection method is also concerned based on k-means algorithm.

Chapter 4 proposed visual perception modelling method based on multi-excitation K-mean clustering and non-negative sparse coding. Multi-firing K-means and non-negative sparse coding are introduced into the HMAX model to establish the biological visual perception model according to the cognitive neurological research results of neural cell characteristics in various regions of the visual cortex of the brain.

Chapter 5 establishes a bio-inspired invariant feature extraction model based on GLoP filter and multi-manifold sparse coding to improve rotation invariance and algorithm efficiency in view of the deficiency of HMAX model which is sensitive to rotation change.

Chapter 6 proposes an incremental pattern association memory model based on the relevant research results of cognitive neuroscience. Leabra learning mechanism is introduced into the model to replace Hebb learning rules, and separate weights are assigned to different pattern categories to achieve incremental learning.

Chapter 7 combines Friston's theory of free energy with the restricted Boltzmann machine to realize the minimization of the free energy of the system based on the RBM model, and to realize the learning and extraction processing of visual information.

Chapter 8 combined the visual attention mechanism with biological visual perception model to realize rapid detection and identification of crop pests based on the research results in Chapter 4 and Chapter 5.

Conclusions summarize the whole book and put forward some suggestions for further works.

The structure diagram of all chapters is shown in Fig. 1.2:

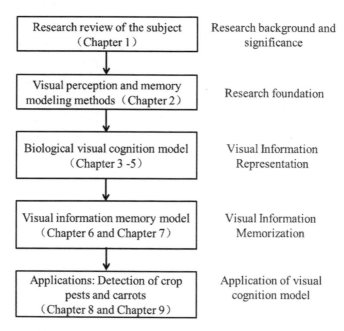

FIGURE 1.2   Structure diagram of each chapter.

1.4 CONCLUSIONS

This chapter first introduces the research background and significance of this book and then summarizes the research status of biological visual perception, human brain memory modelling, and the theory of free energy of human brain cognition. Finally, the main contents of this book are introduced.

REFERENCES

1. Meng, X. 2011. *Research on image recognition technology inspired by biological vision.* PhD dissertation, National University of Defense Technology.
2. Zhang, W. 2014. *Study on feature extraction and its application based on biological visual cognition mechanism.* PhD dissertation, Huazhong University of Science and Technology.
3. Jiang, Y., and Wang Y. 2017. The application of REM memory model in image classification and recognition. *CAAI Transactions on Intelligent Systems* 12(3):310–317.
4. Biederman, I. 1987. Recognition-by-components: a theory of human image understanding. *Psychological Review* 94(2):115–147.
5. Ungerleider, L. G., and Mishkin, M. 1979. The striate projection zone in the superior temporal sulcus of Macaca mulatta: location and topographic organization. *Journal of Comparative Neurology* 188(3):347–366.
6. Itti, L. 2003. Modelling primate visual attention. *Computational Neuroscience: A Comprehensive Approach*: 635–655.
7. Du, X. 2012. *Study on face recognition method inspired by visual perception system.* PhD diss., Chongqing University.
8. Fukushima, K. 1980. Neocognitron: a self-organizing neural network model for a mechanism of pattern recognition unaffected by shift in position. *Biological Cybernetics* 36:193–202.
9. Riesenhuber, M., and Poggio, T. 1999. Hierarchical models of object recognition in cortex. *Nature Neuroscience* 2(11):1019–1025.
10. Serre, T., Wolf, L., Bileschi S., et al. 2007. Robust object recognition with cortex-like mechanisms. *IEEE Transactions on Pattern Analysis & Machine Intelligence* 3:411–426.
11. Maashri, A. A., Debole, M., Cotter M., et al. 2012. Accelerating neuromorphic vision algorithms for recognition. *Proceedings of the 49th Annual Design Automation Conference. ACM*: 579–584.
12. Orchard, G., Martin, J. G., Vogelstein R. J., et al. 2013. Fast neuromimetic object recognition using FPGA outperforms GPU implementations. *IEEE Transactions on Neural Networks & Learning Systems* 24(8):1239–1252.

13. Jiang, L., Wang, Y., and Liu W. 2013. Bio-inspired invariant visual feature representation based on K-SVD and SURF algorithms. *Proceedings of the Fifth International Conference on Internet Multimedia Computing and Service, ACM*:62–65.
14. Mutch, J., and Lowe, D. G. 2008. Object class recognition and localization using sparse features with limited receptive fields. *International Journal of Computer Vision* 80(1):45–57.
15. Pinto, N., DiCarlo, J. J., and Cox, D. D. 2009. How far can you get with a modern face recognition test set using only simple features? *2009 IEEE Conference on Computer Vision and Pattern Recognition (CVPR)*: 2591–2598.
16. Meyers, E., and Wolf, L. 2008. Using biologically inspired features for face processing. *International Journal of Computer Vision* 76(1):93–104.
17. Huang, Y., Huang, K., Tao D., et al. 2011. Enhanced biologically inspired model for object recognition. *IEEE Transactions on Systems Man & Cybernetics Part B Cybernetics A, Publication of the IEEE Systems Man & Cybernetics Society* 41(6):1668–1680.
18. Ghodrati, M., Khaligh-Razavi, S. M., Ebrahimpour, R., et al. 2012. How can selection of biologically inspired features improve the performance of a robust object recognition model. *PloS One* 7(2):e32357.
19. Hormat, A. M., Rostami, V., and Menhaj, M. B. 2013. A robust scene descriptor based on largest singular values for cortex-like mechanisms. *2013 13th Iranian Conference on Fuzzy Systems (IFSC)*: 1–6.
20. Lu, Y. F., Zhang, H. Z., Kang, T. K., et al. 2014. Extended biologically inspired model for object recognition based on oriented Gaussian–Hermite moment. *Neurocomputing* 139:189–201.
21. Tang, T., and Qiao, H. 2014. Improving invariance in visual classification with biologically inspired mechanism. *Neurocomputing* 133:328–341.
22. Eliasi, M., Yaghoubi, Z., and Eliasi A. 2011. Intermediate layer optimization of HMAX model for face recognition. *2011 IEEE International Conference on Computer Applications and Industrial Electronics (ICCAIE)*:432–436.
23. Dura-Bernal, S., Wennekers, T., and Denham S. L. 2012. Top-down feedback in an HMAX-like cortical model of object perception based on hierarchical Bayesian networks and belief propagation. *PloS One* 7(11):e48216.
24. Alameer, A., Ghazaei, G., Degenaar, P., et al. 2016. Object recognition with an elastic net-regularized hierarchical MAX model of the visual cortex. *IEEE Signal Processing Letters* 23(8):1062–1066.
25. Wallis, G., and Rolls, E. T. 1997. Invariant face and object recognition in the visual system. *Progress in Neurobiology* 51(2):167–194.
26. Rolls, E. T., and Stringer, S. M. 2006. Invariant visual object recognition: a model, with lighting invariance. *Journal of Physiology* 100(1-3):43–62.

27. Robinson, L. 2015. *Invariant object recognition: biologically plausible and machine learning approaches*. PhD dissertation, University of Warwick.
28. Rolls, E. T. 2012. Invariant visual object and face recognition: neural and computational Bases, and a model, VisNet. *Frontiers in Computational Neuroscience* 6(35):1–70.
29. Földiák, P. 1991. Learning invariance from transformation sequences. *Neural Computation* 3(2):194–200.
30. Hilton, G., and Salakhutdinov, R. R. 2006. Reducing the dimensionality of data with neural network. *Science* 5786(313): 504–507.
31. Hinton, G. E., Osindero, S., and Teh Y. W. 2006. A fast learning algorithm for deep belief nets. *Neural Computation* 18(7):1527–1554.
32. Sun, Z., Lu, C., Shi, Z., et al. 2016. Research and development of deep learning. *Computer Science* 43(2):1–8.
33. LeCun, Y., Boser, B. E., Denker, J. S., et al. 1990. Handwritten digit recognition with a back-propagation network. *Advances in Neural Information Processing Systems 2* 396–404.
34. Dahl, J. V., Koch, K. C., Kleinhans, E., et al. 2010. Convolutional networks and applications in vision. *IEEE International Symposium on Circuits and Systems* 253–256.
35. LeCun, Y., Jackel, L. D., Bottou, L., et al. 1995. Learning algorithms for classification: A comparison on handwritten digit recognition. *Neural Networks: The Statistical Mechanics Perspective* 261–276.
36. Krizhevsky, A., Sutskever, I., and Hinton, G. E. 2012. ImageNet classification with deep convolutional neural networks. *Communications of the Acm* 60(2):2012.
37. Lecun, Y., Bengio, Y., and Hinton, G. 2015. Deep learning. *Nature*, 521(7553):436.
38. Hubel, D. H. and Wiesel, T. N. 1962. Receptive fields, binocular interaction, and functional architecture in the cat's visual cortex. *Journal of Physiology* 160:106–154.
39. Felleman, D. J. and Essen, D. C. V. 1991. Distributed hierarchical processing in the primate cerebral cortex. *Cerebral Cortex* 1:1–47.
40. Cadieu, C. F. et al. 2014. Deep neural networks rival the representation of primate it cortex for core visual object recognition. *PLoS Computational Biology* 10:e1003963.
41. Fukushima, K. and Miyake, S. 1982. Neocognitron: a new algorithm for pattern recognition tolerant of deformations and shifts in position. *Pattern Recognition* 15:455–469.
42. Waibel, A., Hanazawa, T., Hinton, G. E., Shikano, K. and Lang, K. 1989. Phoneme recognition using time-delay neural networks. *IEEE Trans. Acoustics Speech Signal Process* 37:328–339.
43. Bottou, L., Fogelman-Soulié, F., Blanchet, P. and Lienard, J. 1989. Experiments with time delay networks and dynamic time warping for speaker independent isolated digit recognition. *In Proc. EuroSpeech* 89:537–540.

44. He, K., Zhang, X., Ren S., et al. 2015. Delving deep into rectifiers: Surpassing human-level performance on imagenet classification. *Proceedings of the IEEE International Conference on Computer Vision*:1026–1034.

45. Denton, E., Weston, J., Paluri, M., et al. 2015. User conditional hashtag prediction for images. *Proceedings of the 21th ACM SIGKDD International Conference on Knowledge Discovery and Data Mining*: 1731–1740.

46. Szegedy, C., Liu, W., Jia, Y., et al. 2015. Going deeper with convolutions. *Proceedings of the IEEE Conference on Computer Vision and Pattern Recognition*: 1–9.

47. Toshev, A., and Szegedy, C. 2014. Deeppose: Human pose estimation via deep neural networks. *Proceedings of the IEEE Conference on Computer Vision and Pattern Recognition*: 1653–1660.

48. Sarfraz, M. S., and Stiefelhagen, R. 2015. Deep perceptual mapping for thermal to visible face recognition. *arXiv preprint arXiv:1507.02879*.

49. Ijjina, E. P., and Mohan, C. K. 2014. Facial expression recognition using Kinect depth sensor and convolutional neural networks. *2014 13th International Conference on Machine Learning and Applications (ICMLA)*: 392–396.

50. Byeon, Y. H., and Kwak, K. C. 2014. Facial expression recognition using 3d convolutional neural network. *International Journal of Advanced Computer Science and Applications* 5(12):1–8.

51. Johnson, R., and Zhang, T. 2014. Effective use of word order for text categorization with convolutional neural networks. *arXiv preprint arXiv:1412.1058*.

52. Long, J., and Shelhamer, E., Darrell T. 2015. Fully convolutional networks for semantic segmentation. *Proceedings of the IEEE Conference on Computer Vision and Pattern Recognition*: 3431–3440.

53. O'Reilly, R. C., Munakata, Y., Frank M. J., et al. 2012. *Computational Cognitive Neuroscience*. Wiki book, 1st Edition.

54. Wang, Y., and Chiew, V. 2010. On the cognitive process of human problem solving. *Cognitive Systems Research* 11(1):81–92.

55. Wang, Y. 2009. Formal description of the cognitive process of memorization. *Transactions on Computational Science*. Berlin, Heidelberg: Springer. 81–98.

56. Tulving, E. 1986. Episodic and semantic memory: where should we go from here? *Behavioural and Brain Sciences* 9(3):573–577.

57. Raaijmakers, J. G., and Shiffrin, R. M. 1981. Search of associative memory. *Psychological Review* 88(2):93.

58. Shiffrin, R. M., and Steyvers, M. 1997. A model for recognition memory: REM—retrieving effectively from memory. *Psychonomic Bulletin & Review* 4(2):145–166.

59. Hintzman, D. L. 1988. Judgments of frequency and recognition memory in a multiple-trace memory model. *Psychological Review* 95(4):528.
60. Howard, M. W., and Kahana, M. J. 2002. A distributed representation of temporal context. *Journal of Mathematical Psychology* 46(3):269–299.
61. Rogers, T. T., and McClelland, J. L. 2004. *Semantic cognition: a parallel distributed processing approach.* Cambridge: MIT Press.
62. Burwell, R. D. 2000. The parahippocampal region: corticocortical connectivity. *Annals of the New York Academy of Sciences* 911(1): 25–42.
63. Witter, M. P., Groenewegen, H. J., Da Silva F. H. L., et al. 1989. Functional organization of the extrinsic and intrinsic circuitry of the parahippocampal region. *Progress in Neurobiology* 33(3): 161–253.
64. Eichenbaum, H., Yonelinas, A. P., and Ranganath, C. 2007. The medial temporal lobe and recognition memory. *Annual Review Neuroscience* 30:123–152.
65. Rolls, E. T., Treves, A., and Rolls, E. T. 1998. *Neural Networks and Brain Function.* Oxford, UK: Oxford University Press.
66. Rolls, E. T. 2016. *Cerebral Cortex: Principles of Operation.* Oxford, UK: Oxford University Press.
67. Rolls, E. T., and Kesner R. P. 2006. A computational theory of hippocampal function, and empirical tests of the theory. *Progress in Neurobiology* 79(1): 1–48.
68. McClelland, J. L., McNaughton, B. L., and O'reilly, R. C. 1995. Why there are complementary learning systems in the hippocampus and neocortex: insights from the successes and failures of connectionist models of learning and memory. *Psychological Review* 102(3): 419–457.
69. Norman, K. A., and O'reilly R. C. 2003. Modelling hippocampal and neocortical contributions to recognition memory: a complementary-learning-systems approach. *Psychological Review* 110(4):611–646.
70. Norman, K. A. 2010. How hippocampus and cortex contribute to recognition memory: revisiting the complementary learning systems model. *Hippocampus* 20(11):1217–1227.
71. Bogacz, R., and Brown, M. W. 2003. Comparison of computational models of familiarity discrimination in the perirhinal cortex. *Hippocampus* 13(4): 494–524.
72. Ketz, N., Morkonda, S. G., and O'Reilly, R. C. 2013. Theta doordinated error-driven learning in the hippocampus. *PLoS Computational Biology* 9(6):e1003067.
73. Sederberg, P. B., and Norman K. A. 2010. Learning and memory: computational models. *Encyclopedia of Behavioral Neuroscience* 2: 145–153.
74. Bernard, M., Morrow, J. D., Taylor, S., et al. 2009. Modelling aspects of human memory for scientific study. *Mechanisms of Hippocampal Relational Binding* 1001:87–202.

75. Carpenter, G. A., Grossberg, S., and Rose, D. B. 1991. Fuzzy ART: Fast stable learning and categorization of analog patterns by an adaptive resonance system. *Neural Networks* 4(6):759–771.

76. Adams, F., and Aizawa K. 2001. The bounds of cognition. *Philosophical Psychology* 14(1):43–64.

77. Clark, A. 2013. Whatever next? Predictive brains, situated agents, and the future of cognitive science. *Behavioral and Brain Sciences* 36(3):181–204.

78. Helmholtz, H. V. 1898. Scientific literature: handbuch der physiologischen Optik. *Science* 8:794–796.

79. MacKay, D. M. 1956. The epistemological problem for automata. *Automata Studies* 34: 235–252.

80. Neisser, U. 1976. *Cognition and reality. Principles and implication of cognitive psychology.* San Francisco: WH Freeman and Company.

81. Gregory, R. L. 1980. Perceptions as hypotheses. *Philosophical Transactions of the Royal Society of London B: Biological Sciences* 290(1038):181–197.

82. Rumelhart, D. E., and McClelland J. L., PDP Research Group. 1987. *Parallel Distributed Processing.* Cambridge, MA, USA: MIT Press.

83. Dayan, P., and Hinton, G. E. 1996. Varieties of Helmholtz machine. *Neural Networks* 9(8):1385–1403.

84. Dayan, P., Hinton, G. E., Neal, R. M., et al. 1995. The helmholtz machine. *Neural computation* 7(5):889–904.

85. Rao, R. P. N., and Ballard D. H. 1999. Predictive coding in the visual cortex: a functional interpretation of some extra-classical receptive-field effects. *Nature Neuroscience* 2(1):79–87.

86. Lee, T. S., and Mumford, D. 2003. Hierarchical Bayesian inference in the visual cortex. *JOSA A* 20(7):1434–1448.

87. Friston, K. 2005. A theory of cortical responses. *Philosophical Transactions of the Royal Society of London B: Biological Sciences* 360(1456):815–836.

88. Hinton, G. E., and Zemel, R. S. 1994. Autoencoders, minimum description length and Helmholtz free energy. *Advances in Neural Information processing systems*: 3–10.

89. MacKay, D. J. C. 1995. Free energy minimisation algorithm for decoding and cryptanalysis. *Electronics Letters* 31(6):446–447.

90. Neal, R. M., and Hinton, G. E. 1998. *A view of the EM algorithm that justifies incremental, sparse, and other variants.Learning in graphical models.* Netherlands: Springer. 355–368.

91. Friston, K. J., and Stephan, K. E. 2007. Free-energy and the brain. *Synthese* 159(3):417–458.

92. Friston, K., Kilner, J., and Harrison L. 2006. A free energy principle for the brain. *Journal of Physiology-Paris* 100(1-3):70–87.

93. Friston, K. 2009. The free-energy principle: a rough guide to the brain? *Trends in Cognitive Sciences* 13(7):293–301.

94. Friston, K. 2010. The free-energy principle: a unified brain theory? *Nature Reviews Neuroscience* 11(2):127–138.

95. Friston, K., and Kiebel, S. 2009. Predictive coding under the free-energy principle. *Philosophical Transactions of the Royal Society of London B: Biological Sciences* 364(1521):1211–1221.

## CHAPTER 2

Methods of visual perception and memory modelling

2.1 INTRODUCTION

Eighty percent of the information we get comes from the visual system. Through visual perception and memory systems, the recognition and classification of thousands of objects are completed. This process may seem very simple, but actually, it involves a lot of complex processing. People observe things through their eyes, and then transfer them to the subsequent processing system, which is processed by the visual cortex of the brain through layers of complex processing, and finally transfer them to the memory system for learning and storage, and extract useful information when they need it.

In terms of biological visual perception, with the cooperative efforts of experts in cognitive physiology and neuroscience, we have a deeper understanding of the visual perception mechanism of biological visual systems and the processing process of information in the visual system[1], and have constructed many biological visual perception models. In terms of human brain memory, with the joint efforts of related researchers, people have a more in-depth understanding of human brain activities and have achieved lots of research results, which has laid a solid

DOI: 10.1201/9781003281641-2

theoretical foundation for visual image cognition based on human brain memory mechanism.

Focusing on the issues of how visual information is represented, stored, and extracted in the human brain, this chapter mainly introduces biological visual perception mechanism and visual perception computing model, as well as the neural mechanism, principle, and related memory model of memory.

2.2 MECHANISM AND MODEL OF BIOLOGICAL VISUAL PERCEPTION

2.2.1 Physiological basis of biological visual perception

The process of information processing in the visual system is very complex. The visual system processes information in a hierarchical way through two visual pathways (ventral pathway and dorsal pathway). Visual information mainly reaches the visual cortex of the brain through the ventral visual pathway, forming perception and object recognition; processing behaviour and other spatial information are realized through the dorsal pathway (or where pathway)[2], as shown in Fig. 2.1.

First, people receive external stimulus signals through their eyes, then transfer the signals to LGN, and finally to visual cortex. The visual cortex is divided into primary visual cortex (V1 area), striate cortex, and advanced visual cortex[4]. Among them, the striate cortex includes V2, V3, V4, and other areas, and the high-level visual cortex includes IT, MT, and other areas. As the perception and recognition of objects are mainly responsible for the ventral pathway, the physiological structure and functional characteristics of each region on the ventral visual pathway are briefly introduced below[5].

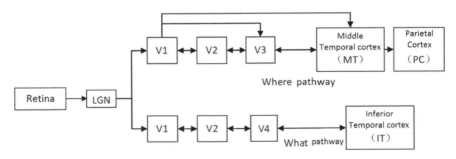

FIGURE 2.1   Two pathways of visual system[3].

As the starting point of the visual pathway, the retina is closely attached to the posterior wall of the eyeball and is responsible for photographic imaging. First, the retina receives external light stimulation and converts it into nerve signal, and then transmits it to the lateral knee body. The lateral geniculate body is a relay structure, its main function is to receive retinal information and transmit to the visual cortex, and receive feedback information from the visual cortex[1].

The primary visual cortex, also known as V1 area, is the most thoroughly understood visual cortex area. Most of the cells in this area are sensitive to long strip stimuli. V1 area cells can be divided into simple cells and complex cells. The receptive field of simple cells is small and rectangular, and the central area is long and narrow; the receptive field of complex cells is larger than that of simple cells and is not sensitive to the position of strip stimulus in the receptive field, but it is selective in the direction of stimulation[6].

V2 region receives input information from V1 region. Neurons in V2 region have certain selectivity to colour, shape, and motion direction, and are sensitive to the stimulation of certain orientation endpoints, such as corner and corner.

V4 region is followed by Region V2. Similar to V1 and V2, V4 cells also show selectivity for location, colour, and direction. However, the difference is that V4 optic nerve cells mainly deal with visual stimuli of moderate complexity, such as simple geometry.

In the high-level visual cortex, the visual features processed by nerve cells become more and more complex. Related studies have shown that some neurons in its area have strong responses to facial images, while some neurons have specific responses to hand shape. This specific reaction shows strong invariance and is not affected by the changes of position, scale, and orientation[3].

2.2.2 HMAX model

HMAX model is a hierarchical maximization model used to simulate the visual cortex of the brain. It has a profound impact on the development of the computational model based on biological visual perception mechanisms. The model is divided into four layers, namely S1 layer, C1 layer, S2 layer, and C2 layer. The model structure is shown in Fig. 2.2. In HMAX model, s layer and C layer appear alternately. S layer is used to simulate the neural characteristics of

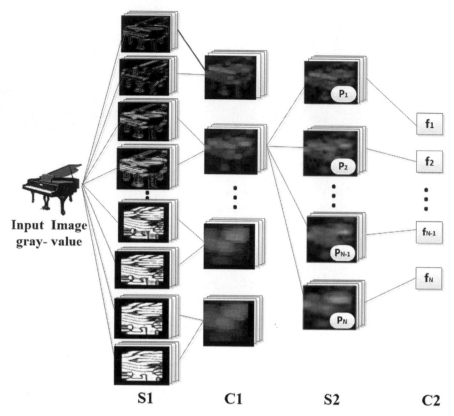

S1	**C1**	**S2**	**C2**

FIGURE 2.2   Overview of HMAX model[7].

simple cells and C layer is used to simulate the neural characteristics of complex cells. The specific implementation of each layer is described in detail below.

S1 layer: S1 layer is used to simulate the neural activities of simple cells in V1 area of primary visual cortex, and a series of Gabor filters are used to simulate the characteristics of receptive field of simple cells[8,9]. The input image is convoluted with a series of Gabor filters with different directions and scales. The definition of Gabor filter function is shown in Equation (2.1):

$$F(x, y) = \exp\left(-\frac{x_0^2 + \gamma^2 y_0^2}{2\sigma^2}\right)\cos\left(\frac{2\pi}{\lambda}x_0\right), \qquad (2.1)$$

$x_0 = x \cos \theta + y \sin \theta$ 且 $y_0 = -x \sin \theta + y \cos \theta$ where γ is the spatial aspect ratio, which determines the shape of Gabor kernel function; θ is the direction, σ is the standard deviation of Gaussian function in Gabor kernel function, and λ is the wavelength parameter of cosine function in Gabor kernel function.

C1 layer: C1 layer is used to simulate the response characteristics of complex cells in V1 region, and the receptive field of complex cells in V1 region is tolerant to changes in position and scale[3]. In other words, the response of a complex unit corresponds to the maximum value of M inputs from S1 layer:

$$r_1 = \max_{j=1\cdots m} x_j \tag{2.2}$$

S2 layer: in S2 layer, the maximum operation of adjacent regions is carried out in all directions of the input image. S2 unit is similar to radial basis function, and the unit response can be obtained by calculating the Euclidean distance between the image block X and the stored feature template. For an image block from C1 layer, the corresponding unit response can be expressed as:

$$r_2 = \exp(-\beta \|X - P_i\|^2) \tag{2.3}$$

where β is the sharpness of tuning the radial basis function and P_i is a feature template obtained by training.

C2 layer: C2 layer is used to simulate the response of neurons in V4 and its area of cerebral visual cortex. The input from S2 layer is maximized globally. The result is an n-dimensional vector, and N is the number of image templates obtained in the training stage.

Training stage: the training stage aims to select a certain number of image templates for S2 layer, and image templates of different sizes are randomly selected from different positions of C1 layer images.

2.3 CONVOLUTIONAL NEURAL NETWORKS

Convolutional neural networks (CNNs) are a variant of multi-layer perceptron (MLP), developed from biologists Huber and Wessel's early research on the visual cortex of cats[10]. There is a complex structure in the cells of the visual cortex, and these cells are very sensitive to the sub-regions of the visual input space, called the receptive field. Convolutional Neural Networks are analogous to traditional ANNs in that they are comprised of neurons that self-optimize through learning. Each neuron will still receive an input and perform an operation (such as a scalar product followed by a non-linear function) – the basis of countless ANNs. From the input raw image vectors to the final output of the class score, the entire network will still express a single perceptive score function (the weight). The last layer will contain loss functions associated with the classes, and all of the regular tips and tricks developed for traditional ANNs still apply[11].

The only notable difference between CNNs and traditional ANNs is that CNNs are primarily used in the field of pattern recognition within images. This allows us to encode image-specific features into the architecture, making the network more suited for image-focused tasks – whilst further reducing the parameters required to set up the model.

CNNs are comprised of three types of layers. These are convolutional layers, pooling layers, and fully-connected layers. When these layers are stacked, a CNN architecture has been formed. A simplified CNN architecture for transport classification is illustrated in Fig. 2.3.

The basic functionality of the example CNN above can be broken down into four key areas.

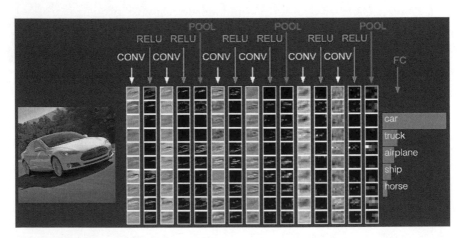

FIGURE 2.3 An simple CNN architecture.

1. The input layer will hold the pixel values of the image.

2. The convolutional layer will determine the output of neurons of which are connected to local regions of the input through the calculation of the scalar product between their weights and the region connected to the input volume. Fig. 2.4 gives an example of convolution operation. The rectified linear unit (commonly shortened to ReLu) aims to apply an 'elementwise' activation function such as sigmoid to the output of the activation produced by the previous layer.

3. The pooling layer will then simply perform downsampling along with the spatial dimensionality of the given input, further reducing the number of parameters within that activation.

4. The fully-connected layers will attempt to produce class scores from the activations, to be used for classification. It is also suggested that ReLu may be used between these layers, to improve performance.

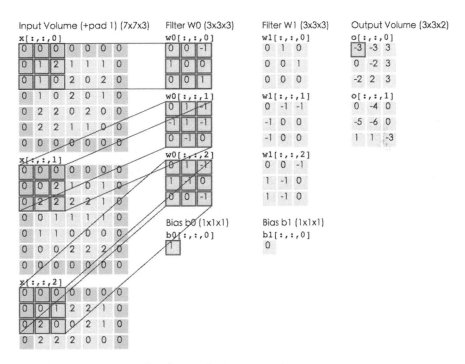

FIGURE 2.4   An example of convolution operation.

LeNet-5[12] proposed by LeCun was the first work of CNN, and its real breakout stage was that AlexNet[13] won the champion of ImageNet classification task in 2012, and the classification accuracy rate was far higher than that achieved by traditional methods. With the success of AlexNet, many works have been proposed to improve its performance[14]. Among them, four representative works are VGGNet[15], GoogleNet[16], ResNet[17], and DenseNet[18].

2.4 NEURAL MECHANISM OF MEMORY

Learning and memory are the basis for human beings to understand the objective world and the advanced functions of the brain. Memory is the maintenance and reproduction of individual experience and learning is the process of acquiring and forming memory[11]. Therefore, it is of great significance to clarify the neural mechanism of learning and memory.

Neurons, also known as nerve cells, are the basic units of the structure and function of the nervous system. Neurons are composed of cell bodies, dendrites, and axons, as shown in Fig. 2.5. Dendrites are multi-rooted and multi-branched processes from the cell body, which are mainly used to receive incoming information. A neuron has multiple dendrites, but only one axon. Neurons transmit information through

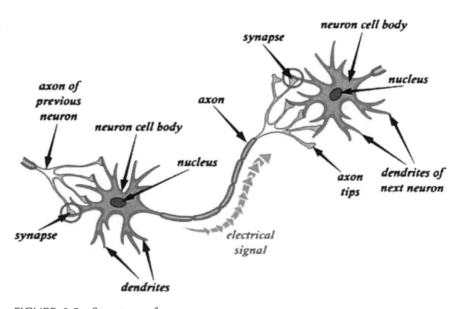

FIGURE. 2.5   Structure of a neuron.

axon terminals located at the end of axons and dendrites of other neurons[19], which are called synapses. The neural basis of learning and memory is the high plasticity of the central nervous system, and synaptic plasticity is the basis of the central nervous system[20].

Memory is formed through long-term changes in synaptic performance, a process known as synaptic plasticity, which is an important physiological basis of behavioural adaptation[20], mainly including short-term synaptic plasticity and long-term synaptic plasticity. Long-term synaptic plasticity can be classified into long-term potentiation (LTP) and long-term depression (LTD), which are regarded as the biological basis of learning and memory activities. In particular, LTP is regarded as one of the main molecular mechanisms for learning and memory.

As early as the end of the 19th century, the famous neurobiologist Cajal thought that the learning process may produce continuous changes in the connection between neurons, which may be the neural basis of memory. At the same time, Tanzi also believes that the synaptic part has the function similar to muscle exercise, which can strengthen the function of the junction part. The basic point of their assumption is synaptic plasticity[21]. In 1949, Hebb[22] put forward the theory of synaptic transmission change. He assumed that some changes of synapses occurred in the learning process, which led to the enhancement of synaptic connections and the improvement of transmission efficiency. Repeated and continuous activities could lead to long-term changes in the connections between neurons. He believes that when the axon of neuron A is enough to stimulate neuron B and activate it repeatedly or continuously, some growth process or metabolic changes will occur in the neuron, thus the efficiency of neuron a will be increased. This process can be simply described as "cells that fire together wire together", which means that when two adjacent neurons discharge together, the connection between them is enhanced, which is the famous Hebb learning theory.

Hebb learning rules can be summarized as follows:

$$\Delta w = xy \qquad (2.4)$$

where Δw represents the change of synaptic weight w and is a function of sending activity x and receiving activity y.

A recent study by Abdou et al.[23] from Fukuyama University in Japan shows that memory is stored in a specific group of neurons called engram cells in the brain. The plasticity of specific synapses represents a specific memory entity, and the synaptic plasticity between specific engram cell combinations is crucial for information storage.

In long-term memory, declarative memory is related to cognitive function, while non declarative memory belongs to unconscious memory, so the research on memory mainly focuses on declarative memory. The results of many years' research have shown that the neuron structure necessary for acquiring declarative memory is located in the medial temporal lobe (MTL). MTL consists of two parts: cortex and subcortical. Cortex includes perinasal cortex, entorhinal cortex (EC), and parahippocampal cortex. The subcutaneous MTL includes the hippocampus and amygdala, and the hippocampal formation is composed of dentate gyrus (DG), CA area, and inferior cortex[24].

The hippocampus is closely connected with other areas of the brain, and the flow of input information between the MTL regions is shown in Fig. 2.6. Most of the information input into the hippocampus first reached the entorhinal cortex and then reached the dentate gyrus through the perforating fibres. After that, CA3/CA1 neurons reached CA1 by themselves and Schreier's fibres, and finally returned to the entorhinal cortex and the inferior supporting area via CA1. There was no direct connection between CA3/CA1 and neocortex. The entorhinal cortex receives input from a large number of cortical regions, mainly from different high-level visual cortex regions, such as temporal cortex it (transnasal cortex) and parietal cortex PC (parahippocampal cortex).

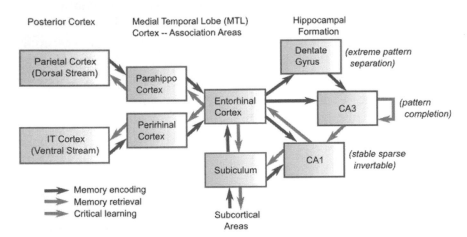

FIGURE 2.6   Information flow in the hippocampal structure[25].

2.5 METHOD OF MEMORY MODELLING

The goal of learning and memory research is to understand how people store and retrieve information based on experience. In order to achieve this goal, many computational models have been proposed to provide formal implementation of memory theory. Memory modelling mainly includes two directions. One goal of memory modelling is to describe the behaviour of participants in the process of memory, that is, from the perspective of cognitive psychology. This kind of model focuses on behaviour reproduction rather than neural data; the other goal is to explain how the brain causes these behaviours, trying to explain neural and behavioural data, that is, model from the perspective of cognitive neurology[26]. Since associative memory has always been a research hotspot in the field of cognitive neuroscience, this section will introduce the related contents of associative memory model.

2.5.1 Memory model based on cognitive psychology

The memory model based on cognitive psychology attempts to describe the mental algorithms supporting recognition and recall without explaining how these algorithms are implemented in the brain. This part mainly introduces two kinds of memory models: REM model[27] and TCM model[28], and semantic memory abstract model Rumelhart model[29].

2.5.1.1 REM model

REM[27] was proposed by Shiffrin and Styvers in 1997. One of the main differences between REM and previous models is that REM realizes Bayesian calculation of likelihood value between cue and a stored memory track while matching calculation is not defined in Bayesian form in previous models.

In REM model, memory is composed of separate images, and each image can be represented as a feature vector $\mathbf{V}$. Feature knowledge is represented by positive integer, and the absence of feature knowledge is represented by 0. A simple geometric distribution of eigenvalues is as follows:

$$P\left[\mathbf{V} = j\right] = (1 - g)^{j-1}g \quad j = 1, 2, \cdots \infty \tag{2.5}$$

Where g is a geometric distribution parameter, different features of a word can be independently generated according to Equation (2.5). High-frequency words have more common features than low-frequency words, and they have higher g value than low-frequency words.

Word learning results are stored in the form of scene images, and the storage vector is an incomplete version of the learned word vector. When learning each word, a piece of information is stored on a feature. If the probability of the value being copied correctly from the learning vector is c, the probability of random selection according to Equation (2.5) is $1-c$.

When extracting, the test vector is matched with the scene images of n words in the vocabulary to get an n-dimensional vector D, where $D_j(j = 1, \ldots, n)$ represents the matching result between the test vector and the jth word. The scene image corresponding to the test word is called s-image (s means the same), and the other scene image is called d-image (d means different).

The key of matching test vector and scene image i, j is to calculate the likelihood ratio λ_j:

$$\lambda_j = \frac{P(D_j|S_j)}{P(D_j|N_j)} \tag{2.6}$$

where S_j represents that I_j is s-image and N_j denotes that I_j is d-image; $P(D_j|S_j)$ is the probability of D_j being observed when I_j is s-image, and $P(D_j|N_j)$ is the probability of D_j being observed when I_j is a d-image.

Then the Bayesian rule is used to calculate the probability that the test word is old (O) (N means new).

$$\Phi = \frac{P(O|D)}{P(N|D)} = \frac{1}{n} \sum_{j=1}^{n} \lambda_j \tag{2.7}$$

where Φ indicates familiarity with a word and can be used to identify decisions. If Φ is greater than 0.5, it is considered that the word has been learned, and the word matches the word corresponding to the maximum value λ_j; otherwise, it is considered as a new word.

2.5.1.2 TCM model

Although REM model provides a mechanism for the memory system to respond to specific cues, it does not describe the behaviour of the memory system when the external cues are not clear, so it needs to generate its own cues for specific memories. In view of this situation, the recovery of psychological context can be used as a means to recall, which plays a key role in memory search theory. Psychological context can be defined as any other information that is actively expressed in the brain when processing a specific stimulus. TCM model[28] is a contextual memory model proposed by Howard and Kahana in 2002. The model uses psychological context to explain the ability of human beings to selectively carry out memory from a specific period of time.

In the learning phase, firstly, the presented item is associated with the current state of the context vector; then, the current state of the context vector is averaged with the semantic features of the newly learned item to update the context. In the recall phase, the current state of the above and below vectors is used as a clue to retrieve the items associated with these context elements in turn. A matrix $\mathbf{M}^{TF}$ is used to relate the psychological context T and items F. Matrix $\mathbf{M}^{TF}$ represents the connection strength between each element t_i in the context and each element f_i in the item. It is composed of a group of outer products: $\mathbf{M}^{TF} = \Sigma f_i t'_i$, where t'_i represents the transposition of vectors t_i, accumulating and covering all the items in the current list.

In the memory stage, the matrix $\mathbf{M}^{TF}$ connects the items to the context T, and the learning rules of the matrix $\mathbf{M}^{TF}$ are as follows:

$$\mathbf{M}_r^{FT} = A_i \mathbf{M}_i^{FT} P_{f_i} + A_i \mathbf{M}_i^{FT} \tilde{P}_{f_i} + B_i t_i f'_i \qquad (2.8)$$

where A_i is attenuation factor, $B_i = A_i/\gamma$, γ is free parameter, P_v is vector projection operator, $\tilde{P}_v = I - P_v$. The context of the current project f_i can be extracted through the matrix $\mathbf{M}^{TF}, t_i^{IN} = \mathbf{M}_i^{FT} f_i$. The current state of the context can be updated by t_i^{IN}, i.e., $t_i = \rho_i t_{i-1} + \beta t_i^{IN}$, where, $0 < \rho_I \leq 1$, β is a free parameter.

2.5.1.3 Rumelhart model

The goal of Rumelhart model[29] is to activate the correct attributes of an item when it is detected with an item and association. The model consists of five layers: project layer, expression layer, correlation layer, hidden layer, and attribute layer. The project layer represents the observed object, the correlation layer represents the context of the observed object,

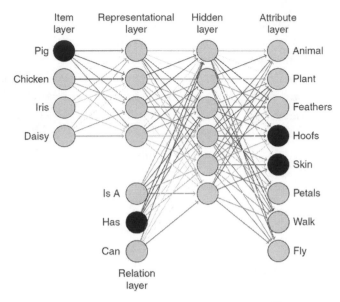

FIGURE 2.7 Simplified diagram of the Rumelhart model[29].

and the attribute layer represents the "invisible" aspect of the predicted object. The basic structure of the model is shown in Fig. 2.7.

When a project unit and an associated unit are activated for detection, the activation propagates forward in the network through the presentation layer and the hidden layer, and finally reaches the attribute layer. The activity pattern of the attribute layer constitutes the prediction result of the model. After the prediction is generated, the network receives the feedback of the project relationship combination actually observed, calculates the error between the predicted attribute and the actual attribute, and then changes the weight to reduce the prediction error. In particular, Rumelhart model uses the back-propagation neural network learning algorithm to adjust the network structure and calculates the values of other units in the network except the project unit to indicate the activity of the unit.

After adequate training, the Rumelhart model learns the context-sensitive mapping between items and attributes. This context sensitivity results from the fact that the activity of the hidden layer is modified by the Association unit. One of the most striking aspects of the model is how to change the internal representation during learning. Items with similar attributes activate similar activity patterns in the presentation layer and hidden layer, and items with different attributes cause different activity patterns.

2.5.2 Memory model based on cognitive neurology

This section introduces a typical model based on cognitive neurology – CLS model, which combines situational memory and semantic memory to form a neural computing framework.

CLS model[30] was proposed by Norman and O'Reilly in 2003, which respectively models hippocampus and neocortex. According to CLS theory, neocortex is a distributed and overlapping system, which is responsible for slow learning of semantic memory, while hippocampus is a sparse and pattern separated system, which is responsible for rapid learning of situational memory. The structure of the CLS model (Fig. 2.8) reflects a broad consensus on the key anatomical and physiological characteristics of the hippocampus and the new cortical subregion, and how these sub-regions contribute to the memory of the whole cortex.

The entorhinal cortex (EC) contains compressed representation of cortical information in other regions. The hippocampal network re-members EC activity patterns by connecting these patterns to a group of units in CA3 region and then returns to EC through CA1. When a pattern is presented, the connections between EC and CA3 active units, between active units within CA3, and between CA3 and CA1 active units are strengthened; on the whole, these synaptic modifications allow the

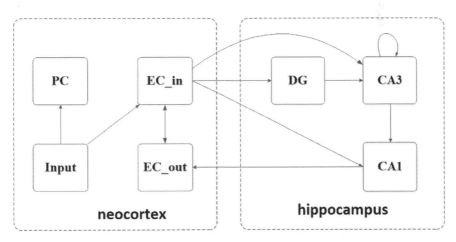

FIGURE 2.8   Architecture of CLS model[31].

network to recall the entire EC pattern stored based on partial cues (i.e., pattern completion). To minimize interference, the network has a built-in bias to assign non-overlapping CA3 representations (i.e., mode separation) to different scenarios. Pattern separation occurs because of the strong feedback suppression in CA3, resulting in sparse representation (i.e., with relatively little neuronal activity). The dentate gyrus (DG) helps to form the pattern separation process, forming a more sparse EC pattern representation, and then enters the region CA3.

The neocortex model is composed of an input layer and a hidden layer. The input layer corresponds to the lower layer of the cortex and is input into the hidden layer in a feedforward manner. The hidden layer includes the perinasal cortex (PC) and EC. The main function of the neocortex model is to extract the statistical laws in the environment. The cortex and hippo region in the model are learned by simple Hebbian learning rules, which strengthen the connection between sending and receiving neurons and weaken the connection between active receiving neurons and inactive sending neurons.

Hippocampal and neocortical networks constitute a recognition memory processing model based on biological stimulation. CLS model assumes that familiarity (i.e., global matching) and recall of specific details contribute to recognition and memory. In CLS model, the hippocampus supports the recall of specific details, but it is not suitable to calculate the global matching between test cues and learning items because different CA3 representations are assigned to different stimuli without considering the similarity between them. The neocortex cannot provide fast enough learning to support recall of details from specific events, but it can calculate familiarity between test cues and learned items.

There are some serious defects in the original form of CLS episodic memory model. In particular, the recognition ability of neocortex network is far lower than that of human recognition memory. This problem is due to the fact that Hebbian learning rules are not wise enough to regulate synaptic strength. Hebbian learning strengthens synapses between active units, even if memories are strong enough to support recall, and weaken the connections between active receiving units and other inactive units, even if those units no longer interfere with memories. This problem can be solved by error-driven learning, which compares the top-down expectation with the perceptual input, and modifies the synapse only when the expectation of the model is not accurate.

2.5.3 Association-based memory model

Combining the research results of neurophysiology and psychophysics with neural computing methods, rolls[32,33] proposed the concepts of pattern association and auto-associative memory, and provided the prototype system of pattern association network and auto-associative memory network.

2.5.3.1 Pattern associative memory network

One of the basic operations of most nervous systems is to learn to associate one stimulus with another that appears at the same time and then to be able to retrieve another given stimulus. The first stimulus may be food, and the other stimulus is its taste. After the association learning, we can know the taste of food when we see it. Pattern association exists at the interface between the output of visual system and learning system, and the association between learning objects, their taste, and tactile stimulation in amygdala. Pattern association also exists in the whole cerebral cortex, which is used to realize top-down influence in attention, emotional influence in memory and visual information processing, visual info

A prototype pattern association network generated according to the basic operation of pattern association is shown in Fig. 2.9. An unconditioned stimulus can activate the activity (or discharge rate) e_i of the i-th neuron.

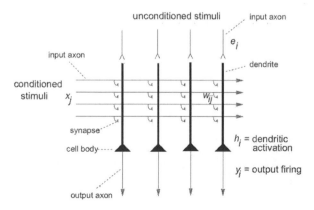

FIGURE 2.9 A pattern association memory network[32].

An unconditioned stimulus can be regarded as a vector of activity e of all neurons, and the excitation rate can also be regarded as a vector of activating y. The conditioned stimulus has the activity (or discharge rate) x_i to the j-th axon, which can also be regarded as a vector x. Unconditioned stimulation is used to generate the activation of output neurons, while conditioned stimulation (equivalent to vector x) is used to generate output neuron dendrites through modifiable synaptic w_{ij}. Synaptic modification satisfies the following rule: when there is a presynaptic activation x_j on an input axon, and there is a postsynaptic activation y_i on neuron i during learning, then the weight w_{ij} between axon and dendrite will increase, which is the famous Hebb learning rule[22].

The excitation rate of each output neuron is determined by the unconditioned stimulus e_i, that is, for each neuron i:

$$y_i = f(e_i) \tag{2.9}$$

This shows that the excitation rate is a function of dendrite activity, and the function f is called activation function, and its form is irrelevant in the learning stage.

Hebb learning rules can be expressed as:

$$\Delta w_{ij} = \alpha y_i x_j \tag{2.10}$$

where Δw_{ij} represents the change of synaptic weight w_{ij}, which is the result of both presynaptic activation x_j and postsynaptic activation y_i. α is the learning rate, which determines the change speed of synaptic weight.

When conditioned stimuli appear on the input axons, the activity h_i on neuron i is the sum of the activities produced by each synaptic strength w_{ij} and each active neuron x_j:

$$h_i = \sum_{j=1}^{C} x_j w_{ij} \tag{2.11}$$

In order to generate cell body activation, it is necessary to convert activity h_i into activation y_i, expressed as:

$$y_i = f(h_i) \tag{2.12}$$

where f is the activation function, where the form of the function becomes more important. Neurons have an activation threshold, which is activated only when the activity of neurons is higher than this threshold.

2.5.3.2 Self-associative memory network

Self-associative memory, or attractor neural network, can be used to store memory. In the network, memory is represented by the pattern of neural activity and stored in the synaptic connections between the network neurons. When a segment of memory is provided, the AAN can extract the appropriate memory from the network, which is called pattern completion. Any different memory can be stored in the network and can be retrieved correctly.

The prototype structure of self-associative memory is shown in Fig. 2.10. The external input e_i is applied to each neuron i through an unmodifiable synapse, thus generating the firing y_i of each neuron. Each output neuron i is connected to other neurons through a modifiable connection weight w_{ij}. The network can effectively associate the output activation vector y with itself during learning. In the recall stage, some output neurons will be activated by the presentation of some external inputs. By repeatedly modifying the synaptic weights, other neurons in y can be activated. This process is repeated many times, and a complete pattern can be recalled.

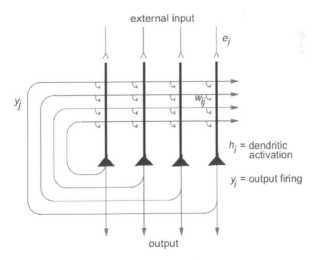

FIGURE 2.10 The architecture of an auto-associative neural network[32].

2.6 CONCLUSIONS

This chapter mainly introduces the neural mechanism of visual perception and human brain memory and the modelling methods of visual perception and memory. This chapter first introduces the mechanism and models of biological visual perception, and then introduces the neural mechanism of memory and method of memory modelling, which lays a theoretical foundation and method support for the progress of the following chapters.

REFERENCES

1. Hu, D. 2012. *Research on image understanding technology based on biological visual perception mechanism.* PhD dissertation, Chengdu: University of Electronic Science and Technology.
2. Serre, T., Kouh, M., Cadieu, C., et al. 2005. A theory of object recognition: computations and circuits in the feedforward path of the ventral stream in primate visual cortex. *Summary of Research on Emotional Handicap Education* 20(7):527–532.
3. Jiang, L. 2014. *Research on image invariance feature representation method inspired by biological visual perception mechanism.* M. S. thesis, China University of Petroleum.
4. Luo, S. 2010. *Visual Information Cognitive Computing Theory.* Beijing: Science Press.
5. Acock, M. 1985. Vision: a computational investigation into the human representation and processing of visual information. by David Marr. *The Modern Schoolman* 62(2):141–142.
6. Du, X. 2012. *Study on face recognition method inspired by visual perception system.* PhD dissertation, Chongqing University.
7. Serre, T., Wolf, L., Bileschi, S., et al. 2007. Robust object recognition with cortex-like mechanisms. *IEEE Transactions on Pattern Analysis & Machine Intelligence* (3):411–426.
8. Hubel, D. H., and Wiesel, T. N. 1962. Receptive fields, binocular interaction and functional architecture in the cat's visual cortex. *The Journal of Physiology* 160(1):106–154.
9. Jones, J. P., and Palmer L. A. 1987. An evaluation of the two-dimensional Gabor filter model of simple receptive fields in cat striate cortex. *Journal of Neurophysiology* 58(6):1233–1258.
10. Hubel, D. H., and Wiesel, T. N. 1968. Receptive fields and functional architecture of monkey striate cortex. *The Journal of Physiology* 195(1): 215–243.
11. O'Shea, K., and Nash, R. 2015. An introduction to convolutional neural networks. Computer Science. Okano, H., Hirano, T., Balaban, E. 2000.

Learning and memory. Proceedings of the National Academy of Sciences 97(23):12403–12404.

12. LeCun, Y., Bottou, L., Bengio, Y., and Haffner, P. 1998. Gradient-based learning applied to document recognition. *Proceedings of the IEEE* 86(11):2278–2324.

13. Krizhevsky, A., Sutskever, I., and Hinton, G. E. 2012. ImageNet classification with deep convolutional neural networks. In: *Advances in Neural Information Processing Systems* pp. 1097–1105.

14. Gu, J., Wang, Z., Kuen, J., Ma, L., Shahroudy, A., et al. 2018. Recent advances in convolutional neural networks. *Pattern Recognition* 77: 354–377.

15. Simonyan, K., and Zisserman, A. 2014. Very deep convolutional networks for large-scale image recognition. *arXiv preprint arXiv*:1409.1556.

16. Szegedy, C., Liu, W., Jia, Y., Sermanet, P., Reed, S., Anguelov, D., and Rabinovich, A. 2015. Going deeper with convolutions. In *Proceedings of the IEEE conference on computer vision and pattern recognition* (pp. 1–9).

17. He, K., Zhang, X., Ren, S., and Sun, J. 2016. Deep residual learning for image recognition. In *Proceedings of the IEEE conference on computer vision and pattern recognition* (pp. 770–778).

18. Iandola, F., Moskewicz, M., Karayev, S., Girshick, R., Darrell, T., and Keutzer, K. 2014. Densenet: Implementing efficient convnet descriptor pyramids. *arXiv preprint arXiv*:1404.1869.

19. Qi Y. 2012. *Research on robust moving target extraction and tracking method based on human memory mechanism.* PhD dissertation, China University of Petroleum (East China).

20. Xiu, D., and Xue, H. 2013. Research progress on neural mechanism of learning and memory. *Bulletin of Biology* 48(8):1–3.

21. Wu, F., and Xiao, X. 1991. Learning, memory and synaptic mechanisms in the brain. *Chinese Journal of Nature* (10):746–751.

22. Hebb, D. O. 1949. *The Organization of Behavior: A Neuropsychological theory.* New York: Science Editions.

23. Abdou, K., Shehata, M., Choko, K., et al. 2018. Synapse-specific representation of the identity of overlapping memory engrams. *Science* 360(6394):1227–1231.

24. Rutishauser, U. 2008. *Learning and representation of declarative memories by single neurons in the human brain.* PhD dissertation, California Institute of Technology.

25. O'Reilly, R. C. Munakata, Y., Frank, M. J., et al. 2012. *Computational Cognitive Neuroscience.* Wiki Book, 1st Edition.

26. Sederberg, P. B., and Norman K. A. 2010. Learning and memory: computational models. *Encyclopedia of Behavioral Neuroscience* 2: 145–153.

27. Shiffrin, R. M., and Steyvers, M. 1997. A model for recognition memory: REM—retrieving effectively from memory. *Psychonomic Bulletin & Review* 4(2):145–166.

28. Howard, M. W., and Kahana, M. J. 2002. A distributed representation of temporal context. *Journal of Mathematical Psychology* 46(3): 269–299.
29. Rogers, T. T., and McClelland, J. L. 2004. *Semantic Cognition: A Parallel Distributed Processing Approach*. Cambridge: MIT Press.
30. Norman, K. A., and O'reilly, R. C. 2003. Modelling hippocampal and neocortical contributions to recognition memory: a complementary-learning-systems approach. *Psychological Review* 110(4): 611–646.
31. Norman, K. A. 2010. How hippocampus and cortex contribute to recognition memory: revisiting the complementary learning systems model. *Hippocampus* 20(11): 1217–1227.
32. Rolls, E. T. 2016. *Cerebral Cortex: Principles of Operation*. Oxford, UK: Oxford University Press.
33. Rolls, E. T., and Kesner, R. P. 2006. A computational theory of hippocampal function, and empirical tests of the theory. *Progress in Neurobiology* 79(1): 1–48.

## CHAPTER 3

Bio-inspired model for object recognition based on histogram of oriented gradients

3.1 INTRODUCTION

A human can easily recognize objects despite the great variations in size, position, view, and background[1]. In recent years, many computational models try to emulate the recognition mechanism of the remarkable visual system of humans, inspired by cognitive neuroscience research achievements. Among these models, one extraordinary achievement was the HMAX model proposed by Riesenhuber and Poggio, which tried to model the recognition behaviour of the visual cortex[2]. Serre et al. significantly improved the HMAX model and proposed the standard HMAX[3].

Although the standard HMAX model shows excellent ability for object recognition, it has two limitations: one is its processing speed and the other is its poor invariance to illumination and rotation[3]. Over the past few years, many efforts have been made to improve the HMAX model. Mahdavi Hormat et al. used largest singular values to improve the invariance to rotational change[4]. However, it is a time-consuming work to

DOI: 10.1201/9781003281641-3

compute the singular values. Jiang et al. combined the HMAX model with SURF algorithm to improve the rotational invariance[5]. But very few templates might be extracted with small number of training images. Additionally, some methods originated from HMAX for face recognition and proved to be comparable and sometimes superior to some other prevalent representations[6-8].

To improve the processing speed and robust to illumination and transformations, we integrate HOG algorithm[9] into the standard HMAX model. HOG algorithm maintains a few key advantages over other feature extraction methods and captures local representation with invariance to illumination change, local geometric transformations. Meanwhile, a normalized dot-product operation is used to compute the S2 unit response instead of Euclidean distance. Furthermore, an effective prototype selection method based on k-means algorithm[10] is also concerned[11-13].

3.2 RELATED WORK

3.2.1 HMAX model

The standard HMAX is a biologically inspired model with four layers: S1, C1, S2, and C2 (as shown in Fig. 3.1).

3.2.2 HOG algorithm

HOG algorithm was proposed by Dalal and Triggs for pedestrian detection[9]. The basic idea behind HOG descriptor is that local object shape and appearance can be represented by the distribution of intensity gradients. As depicted in Fig. 3.2, the image is divided into small connected regions, called cells, and a histogram of gradient orientations is computed for the pixels within each cell. The collection of these histograms then forms the descriptors (see[9] for details). For better invariance to illumination or shadowing changes, local histograms are contrast normalized across a larger region, called block.

3.3 HOG-HMAX MODEL

The model proposed in this book denoted as HOG-HMAX, involves three modifications to the HMAX model. Fig. 3.3 presents the overall flowchart of our model in which the modifications are highlighted in red.

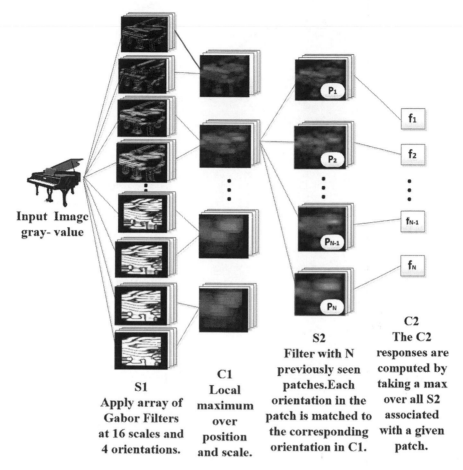

FIGURE 3.1   Overview of standard HMAX model.

3.3.1 S1 layer

S1 layer receives grey images as input and convolves with a battery of Gabor filters which are defined by:

$$F(x, y) = \exp\left(-\frac{x_0^2 + r^2 y_0^2}{2\sigma^2}\right)\cos\left(\frac{2\pi}{\lambda}x_0\right)$$

$$x_0 = x \cos\theta + y \sin\theta,$$ \hspace{2cm} (3.1)

$$y_0 = -x \sin\theta + y \cos\theta$$

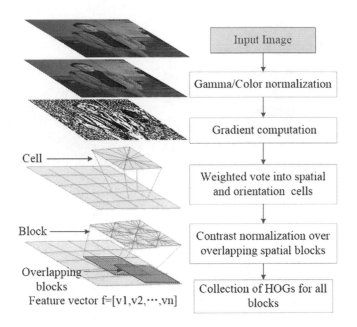

FIGURE 3.2   Whole process of HOG algorithm.

where λ, θ, σ, and r represent respectively the wavelength, orientation, standard deviation, as well as the spatial aspect ratio.

According to Eliasi and Yaghoubi[8], the first four bands of Gabor filters in S1 layer are more efficient than using all bands in object recognition. So we only use the first four bands to increase the efficiency.

3.3.2 C1 layer

C1 layer first pools over afferent S1 units with the same orientation and from the same scale band[3]. And then, HOG descriptors[9] are extracted from the subsampled images. Each image is divided into small cells, which will be grouped into larger connected blocks. Thus the HOG descriptor is the vector of the normalized histograms over block regions. Column of the HOG descriptor corresponds to the local feature of a block region and column number indicates number of all spatial blocks.

Feature vector of HOG-based image patches contains fewer values than that of HMAX. For an image patch of size 8×8, it contains 64 values. But for HOG feature, supposing each block contains 2×2 cells, the feature vector of an image block consists of only 36 values. Feature vector dimension of HOG descriptor is independent of size of cells and only depends on how many cells are grouped into a block.

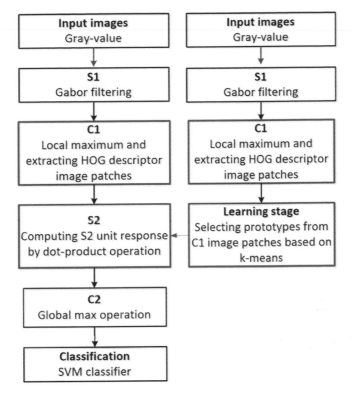

FIGURE 3.3   Flowchart of the proposed model.

3.3.3 S2 layer

The second modification is that each S2 unit response is calculated by a normalized inner product. In consideration of the essential properties of synapses, it is proved that a single neuron can perform high dimensional (10^3–10^4) inner product operations between input and the stored vectors [11]. So inner product [12] is more appropriate to compute the simple cell response than Euclidean distance. For an image patch X at a particular band scale and a stored prototype P_i, the corresponding S2 unit response r can be computed by:

$$r = |X||P_i| \cos\ <X, P_i> \qquad (3.2)$$

where |X| represents the length of vector X and $<X, P_i>$ $<X, P_i>$ indicates the angle between the two vectors.

3.3.4 C2 layer

The C2 response is achieved by a global maximum over all scales and positions for each S2 unit as in standard HMAX model, which will result in a vector of N values, where N corresponds to the number of prototypes.

3.3.5 Prototypes learning stage

The final modification lies in the learning stage which aims to select a set of N prototypes for computing S2 unit response. First we extract all available HOG-based image patches from a target set of images. Then we use k-means based selection criterion to select N informative prototypes. The steps can be described as below.

Step 1 Initialization. Set the number of desired clusters with N, and then N image patches are randomly chosen to be the cluster centres.

Step 2 Use k-means algorithm to group all the available image patches into N clusters.

Step 3 Choose the final N cluster centres as the prototypes.

3.4 RESULTS

To evaluate the performance of the HOG-HMAX model, we conduct recognition tasks on three benchmark databases. After the C2 features are extracted, they are fed to the SVM classifier[13] to perform binary recognition task. All the experiments are run for 10 times and average the results. Four orientations and the first four bands of Gabor filters are used. In HOG algorithm, cell size is same as the spatial pooling grid of corresponding scale band. The parameters of C1 layer, C2 layer and HOG algorithm are shown in Table 3.1.

3.4.1 Caltech5 dataset

The Caltech5 dataset consists of five object categories: airplanes, rear cars, frontal faces, leaves, and motorcycles. We compare our model with the standard HMAX and the SIFT algorithm[14] at the features' level. The Caltech5 dataset is used as the positive dataset and background is used as negative dataset. As described in[3], we randomly select 40 images from

TABLE 3.1   Summary of the S1 layer, C1 layer, and HOG algorithm parameters

C1 layer			S1 layer			HOG algorithm		
Scale band S	Spatial pooling grid ($Ns \times Ns$)	Overlap Δs	Filter size s	Gabor σ	Gabor λ	Bin number	Angle	Cell size
Band 1	8×8	4	7×7	2.8	3.5	9	360	8
			9×9	3.6	4.6			
Band 2	10×10	5	11×11	4.5	5.6	9	360	10
			13×13	5.4	6.8			
Band 3	12×12	6	15×15	6.3	7.9	9	360	12
			17×17	7.3	9.1			
Band 4	14×14	7	19×19	8.2	10.3	9	360	14
			21×21	9.2	11.5			

each category as positive training samples and 50 images as negative training samples. For testing, we randomly select 50 positive testing samples and 50 negative testing samples from the remaining images. All images are resized to 140 pixels in height (width is resized accordingly). Fig. 3.4 shows some sampling images of each category. There are great differences in appearance within each category.

To study the contribution of the number of features on the recognition performance, we extract different numbers of features (5, 10, 50, 100, 200, and 500) for each image. We choose the classification rate for various numbers of features as the evaluation criterion. Fig. 3.5 shows the comparison among the SIFT, the standard HMAX, and HOG-HMAX under different numbers of features. In general, HOG-HMAX outperforms the HMAX and SIFT algorithm. The recognition ability is different with different categories. Relatively speaking, our method shows good recognition ability and significantly outperformed the standard HMAX for airplanes, motorcycles, cars, and leaves. But for faces, it does not achieve good results, and the recognition performance is roughly equivalent to the HMAX model. The classification rate increases with more features, but a satisfactory performance has already been obtained with 200 features. So in consideration of computational time and recognition performance, we extract 200 C2 features in the following experiments on Caltech101 dataset and Caltech256 dataset.

FIGURE 3.4 Sampling images of Caltech5 dataset: airplanes, cars, faces, leaves, motorcycles, and background.

To investigate the influence of the number of training examples on recognition performance, we choose variable number of positive training samples (1, 3, 6, 15, 30, and 40) from each object category and 50 negative training samples. For testing, we choose 50 positive samples and 50 negative samples from the remaining images. Fig. 3.6 shows the comparison of HOG-HMAX with HOG and standard HMAX on the Caltech5 dataset with different numbers of training images. For all three methods, with the number of training images increasing, the classification performance can be improved accordingly. The classification can obtain a reasonable result with 15 training images. In general, the performance of HOG-HMAX is superior to the HOG and standard HMAX model with different numbers of training examples.

In this experiment, the recognition performance of faces category has not been improved so obviously compared with other four categories. Dalal and Triggs found that the optimal parameters were 3×3 cell blocks of 6×6 pixel cells with 9 histogram channels in human detection[9]. While in our experiment, the HOG parameters are 2×2 cell blocks with

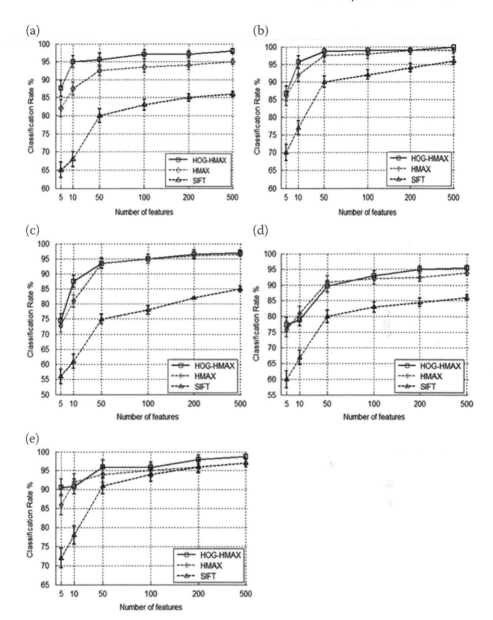

FIGURE 3.5 Comparison of HOG-HMAX with SIFT and standard HMAX on the Caltech5 dataset: (a) airplanes, (b) cars, (c) faces, (d) leaves, and (e) motorcycles.

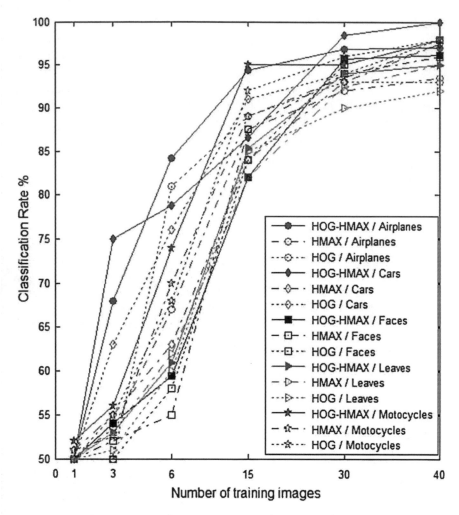

FIGURE 3.6 Comparison of HOG-HMAX with HOG and standard HMAX on the CalTech5 database with different numbers of training images.

8–12 cell size, which is likely to cause face recognition not achieving the optimal results.

3.4.2 Caltech101 dataset

The Caltech101 dataset consists of 101 object categories and a background category (see Li et al.[15] for details). We select six object categories (bonsai, pizza, pyramid, sunflower, Windsor chair, and leopards) as positive dataset and background as negative dataset. Twenty images

FIGURE 3.7 Sampling images of Caltech101 dataset: pyramid, bonsai, sunflower, leopards, pizza, and Windsor chair.

are randomly selected from each category as positive training examples and 30 images as negative training examples. For testing, we use 30 positive examples and 30 negative examples. Some sampling images are shown in Fig. 3.7. All images are resized to 140 pixels in height and width is resized accordingly. The experiment is conducted on PC computer (CPU: Pentium(R) Dual-Core CPU E5800 @ 3.20GHZ, RAM: 2G; OS: Win7).

Table 3.2 presents the experimental results of SSMF[5], HMAX, and HOG-HMAX on Caltech101 dataset including classification rate and processing time. The results show that our method outperforms the HMAX and SSMF model not only in classification rate but also in processing time. The standard HMAX model spends about three times as much as our method.

TABLE 3.2   Comparison of SSMF, HMAX, and HOG-HMAX on six categories from Caltech101 dataset

Categories	SSMF		HMAX		HOG-HMAX	
	[a]CR(%)	Time(s)	CR(%)	Time(s)	CR(%)	Time(s)
pyramid	82.6	94.55	84.0	186.91	**89.2**	**56.72**
bonsai	75.8	87.38	83.3	169.47	**88.3**	**50.67**
sunflower	75.8	89.55	75.3	163.35	**80.0**	**56.46**
leopards	80.4	94.89	81.3	181.34	**84.2**	**60.64**
pizza	82.5	91.74	81.7	171.75	**92.5**	**53.62**
Windsor chair	93.3	83.28	89.3	149.06	**95.0**	**46.22**

Note
[a] CR: Classification rate.

3.4.3 Caltech256 dataset

To evaluate the recognition performance of our model under more complex conditions, we conduct experiment on the challenging Caltech256 dataset. In this experiment, we choose five object categories (French horn, tennis ball, tomato, t-shirt, and watermelon) as positive datasets and clutter as negative datasets. We use 20 positive training examples and 30 negative training examples. For testing, we use 30 positive examples and 30 negative examples.

Table 3.3 demonstrates performance comparison of HOG-HMAX and standard HMAX on Caltech256 dataset and Fig. 3.8 depicts the ROC curves[16] averaged over 10 independent trials. In this experiment, even though the objects are more complex than those of Caltech101, the HOG-HMAX consistently outperforms the standard HMAX in general.

TABLE 3.3   Comparison of HMAX and HOG-HMAX on Caltech256 dataset

Categories	HMAX		HOG-HMAX	
	[b]EER	[c]AUC	EER	AUC
Tennis ball	72.1	76.4	**76.3**	**83.4**
Tomato	78.4	84.1	**82.1**	**86.9**
T-shirt	79.2	85.0	**85.0**	**92.0**
Watermelon	80.0	88.2	**80.3**	**90.1**
French horn	84.3	92.8	**92.3**	**97.1**

Notes
[b] EER: Detection rate at equal-error-rate of the ROC curve.
[c] AUC: Area under the ROC curve.

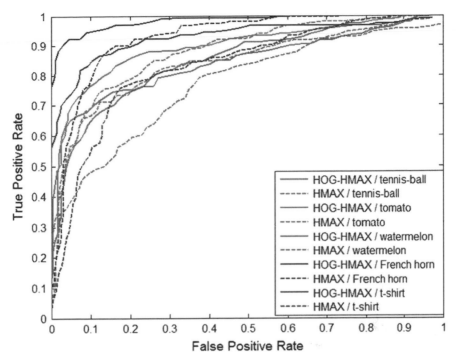

FIGURE 3.8 ROC curves of HOG-HMAX compared with HMAX on Caltech256 dataset.

3.5 CONCLUSIONS

The major contribution of this chapter is to integrate HOG descriptors into the HMAX model, a well-known biologically inspired computational model, to create a simple but effective model called HOG-HMAX. Combined with HOG descriptors, higher-level information can be learned along with the hierarchical structure of the model. The inner product is used to compute the S2 response due to performance and efficiency consideration.

We have conducted experiments on three benchmark databases to evaluate the performance of the proposed model. Results show that the HOG-HMAX model can not only learn an invariant representation of objects but also own some selectivity at the same time. In contrast to the standard HMAX model, its tolerance to rotational and photometric change has been improved to some degree, especially local transformation. But it still has limitations in global rotation conditions. Applying rotation-invariant HOG Descriptors[17] may give further improvement in recognition performance.

We have taken some effort to improve the processing speed, such as selecting fewer Gabor filter bands, using efficient inner product operation to compute the S2 response, and extracting smaller size of feature vector. Experiment on Caltech101 dataset shows that HOG-HMAX outperforms the HMAX and SSMF model both in recognition performance and time efficiency. So the proposed model can meet the requirements of practical applications much better.

Model parameters play a critical role in recognition performance. Tuning appropriate values to these parameters can achieve better results. But for different object recognition tasks, the optimal value of parameters may be different. In this article, we use the same parameters for all recognition tasks, which leads to some categories not achieving their optimal performance, for example, faces in Caltech5. So further improvement can perhaps possibly be achieved by tuning some of these parameters through learning.

REFERENCES

1. Li, H., Li, H., Wei, Y., Tang, Y., and Wang, Q. 2014. Sparse-based neural response for image classification. *Neurocomputing* 144:198–207.
2. Riesenhuber, M., and Poggio, T. 1999. Hierarchical models of object recognition in cortex. *Nature Neuroscience* 2(11):1019–1025.
3. Serre, T., Wolf, L., Bileschi, S., Riesenhuber, M., Poggio, T. 2007. Robust object recognition with cortex-like mechanisms. *IEEE Transactions on Pattern Analysis and Machine Intelligence* 29(3): 411–426.
4. Mahdavi Hormat, A., Rostami, V., Menhaj, M. B. 2013. A robust scene descriptor based on largest singular values for cortex-like mechanisms. *2013 13th Iranian Conference on Fuzzy Systems (IFSC)*:1–6.
5. Jiang, L., Wang, Y., Liu, W. 2013. Bio-inspired invariant visual feature representation based on K-SVD and SURF algorithms. *International Conference on Internet Multimedia Computing and Service (ICIMCS2013)*: 62–65.
6. Pinto, Y. N., Di Carlo, J. J., Cox, D. D. 2009. How far can you get with a modern face recognition test set using only simple features? *IEEE Conference on Computer Vision and Pattern Recognition*: 2591–2598.
7. Meyers, E., Wolf, L. 2008. Using biologically inspired features for face processing. *International Journal of Computer Vision* 76(1): 93–104.
8. Eliasi, M., Yaghoubi, Z. 2011. Intermediate layer optimization of HMAX model for face recognition. *2011 International Conference on Computer Applications and Industrial Electronics (ICCAIE 2011)*:432–436.
9. Dalal, N., and Triggs, B. 2005. Histograms of oriented gradients for human detection. *Computer Vision and Pattern Recognition* 1:886–893.

10. Jain, A. K. 2010. Data clustering: 50 years beyond K-means. *Pattern Recognition Letters* 31(8): 651–666.
11. Deng, L., and Wang, Y. 2016. Bio-inspired model for object recognition based on histogram of oriented gradients. *2016 12th World Congress on Intelligent Control and Automation (WCICA)*: 3277–3282
12. Mcculloch, W. S., and Pitts, W. 1943. A logical calculus of the ideas immanent in the nervous activity. *The Bulletin of Mathematical Biophysics* 5(4): 115–133.
13. Noble, W. S. 2006. What is a support vector machine? *Nature Biotechnology* 24(12): 1565–1567.
14. Lowe, D. G. 1999. Object recognition from local scale-invariant features. *Proc. Int'l Conf. Computer Vision*: 1150–1157.
15. Li, F. F., Fergus, R., Perona, P. 2007. Learning generative visual models from few training examples: an incremental Bayesian approach tested on 101 object categories. *Computer Vision and Image Understanding* 106(1): 59–70.
16. Ho, P. S., Mo, G. J., Chan-Hee, J. 2004. Receiver operating characteristic (ROC) curve: practical review for radiologists. *Korean Journal of Radiology* 5(1): 11–18.
17. Liu, K., Skibbe, H., Schmidt, T., Blein, T., Palme, K., Brox, T., Ronneberger, O. 2014. Rotation-invariant HOG descriptors using Fourier analysis in polar and spherical coordinates. *International Journal of Computer Vision* 106(3): 342–364.

## CHAPTER 4

Modelling object recognition in visual cortex using multiple firing K-means and non-negative sparse coding

4.1 INTRODUCTION

Humans can easily recognize objects despite the tremendous variation in size, position, lighting condition, viewpoint, and other sources of disturbance and noise[1]. This remarkable ability is mainly supported by the ventral visual stream[2] (as shown in Fig. 4.1). The ventral stream starts from the lateral geniculate nucleus (LGN), then goes through the primary visual cortex (V1), and travels through V2 and V4 to the inferior temporal (IT) area[3]. Neurons in different areas have different properties. For instance, neurons in V1 areas are sensitive to bars, and neurons in V2 areas are sensitive to corners[4]. While neurons in V4 areas are

DOI: 10.1201/9781003281641-4

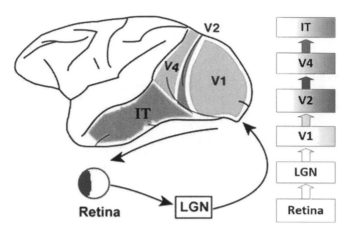

FIGURE 4.1 Ventral visual stream located in the brain of macaque monkey and visual information flow from the retina[3].

responsive to boundary information[5], IT neurons respond to complicated patterns[2] and can directly support invariant object recognition[6,7].

Motivated by the neural properties of ventral visual stream, many efforts have been taken to simulate the process of object recognition. Significant progresses have been made in interpreting the properties of neurons in lower level of visual cortex such as V1, which are well specified by Gabor-like detectors[8]. But relatively few efforts have been made to reproduce the neural properties beyond V1[9]. The famous hierarchical model HMAX[10] has shown its outstanding ability to interpret some properties of V1 and V4 neurons[11]. It comprises two S layers and two C layers. The S1 and C1 layer interpret response properties of the cells in V1 area, and C2 layer can reproduce some neural properties in V4 and IT areas[12]. HMAX model is a promising model for mimicking the mechanism of the visual cortex of primates[12] and is capable of duplicating some neural properties of V1 and V4 areas, but some limits still exist in the following respects. First, it doesn't account for the properties of V2 neurons. Secondly, the templates are selected randomly during the learning stage. Thus, its performance reliability relies on large numbers of templates[13], which will increase the computational cost. Last, though C2 layer can reproduce some neural properties in the V4 and IT areas, it doesn't reflect the properties of sparse firing of the V4 neurons.

In the past few years, some attempts have been made in modelling response properties of V2 neurons[14–17]. Jeremy et al. proposed a model for middle ventral processing and provided a new functional explanation of V2 neurons[14]. Lee et al. found that a two-layer DBN with sparse constraints on each layer could reproduce some neural properties of V1 and V2 areas[16]. However, the sparse DBN is extremely abstract and too complicated for biological systems[17]. Inspired by the sparse DBN, Hu et al. proposed an effective model for V2 neurons originating from the k-means clustering algorithm, which is more efficient and biologically plausible than the sparse DBN[9].

As an unsupervised learning method, sparse coding plays an important role in representing objects of external world[18] and can well account for the receptive fields of V1 cells[19–21]. A recent study showed that sparse coding could better interpret the response properties of V4 neurons[22]. So, in this chapter, we incorporate sparse coding into the HMAX model to emulate the activity of V4 cells. To reduce the information loss of sparse coding[23], non-negative sparse coding (NNSC) is applied for reproducing the V4 neural responses. In addition, we propose an effective template selection method during the learning stage[24].

The rest of this chapter is organized as follows. We briefly introduce some related work including standard model of visual cortex and non-negative sparse coding in Section 4.2. In Section 4.3, we elaborate the proposed model and show how to model the V2 and V4 neural responses. Section 4.4 provides empirical results on three public image databases. Finally, we give conclusions and directions for future works in Section 4.5.

4.2 RELATED WORK

In what follows, we briefly review some methods which are closely related to the proposed method in this section.

4.2.1 The HMAX model

The HMAX model proposed in[12] attempts to model the ventral visual pathway from V1 to higher areas of ventral visual stream[25] and comprises four layers, as illustrated in Fig. 4.2.

S1 layer: The input image is convolved with a battery of Gabor filters with different scales and orientations[12]. Gabor filters are employed to imitate V1 simple cells and described as below:

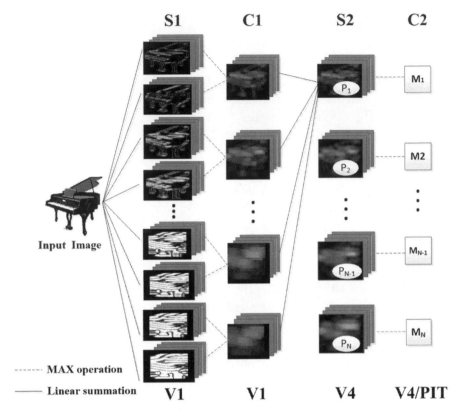

FIGURE 4.2   Structure of the standard HMAX model.

$$f(x, y) = \exp\left(-\frac{x_0^2 + r^2 y_0^2}{2\sigma^2}\right) \times \cos\left(\frac{2\pi x_0}{\lambda}\right) \qquad (4.1)$$

Where,

$$x_0 = x \cos \theta + y \sin \theta, \; y_0 = -x \sin \theta + y \cos \theta$$

r is the aspect ratio and λ is the wavelength.

C1 layer: This layer is expected to emulate the complex cells in V1 area. C1 units pool over S1 units from the same scale band and with the same orientation[12].

S2 layer: S2 layer responses depend on the Euclidean distance between a stored template and an image patch. The S2 unit response is calculated by:

$$r = \exp(-\beta \|X - P_i\|^2) \tag{4.2}$$

where β indicates the tuning sharpness and P_i presents one of the N templates which are randomly extracted during the learning stage.

C2 layer: C2 responses are calculated by a global maximum operation over all positions and scales[12]. From the biological viewpoint, the C2 layer models the V4 neurons with high invariance and selectiveness[26].

4.2.2 Non-negative sparse coding (NNSC)

Assume $X = [x_1, x_2, \cdots, x_N] (x_i \in R^{D \times 1})$ is the set of N vectors of D dimensions. The sparse coding problem can be defined by:

$$\min \|X - WS\|^2 + \lambda \|S\|_1 \tag{4.3}$$

where $W = [w_1, w_2, \cdots, w_K] (w_i \in R^{D \times 1})$ is the basis matrix, $S = [s_1, s_2, \cdots, s_N] (s_i \in R^{K \times 1})$ is the contribution of each basis vector[27] and λ is the regularization parameter.

However, there is a problem with sparse coding. The standard sparse coding is unrealistic to emulate the properties of V1 simple cell for the fact that each neuron can be either positively or negatively active. Thus, non-negative sparse coding was proposed for biological modelling. NNSC tries to solve the following minimization problem:

$$C(A, S) = \|X - WS\| + \lambda \|S\|_1 \tag{4.4}$$

$$s.\, t.\; W_{ij} \geq 0,\; S_{ij} \geq 0,\; \|w_i\| = 1,\; \forall\, i, j$$

There are a large number of approaches to solve sparse coding problem[28-37]. The algorithm used in this book follows[35], which provides an efficient implementation for non-negative sparse coding.

4.3 OVERVIEW OF THE PROPOSED SPARSE-HMAX MODEL

4.3.1 Structure of the proposed method

We extend the standard HMAX model in some biologically plausible ways. We add a hidden layer between C1 and S2 layers and construct five-layer architecture. Meanwhile, NNSC is integrated into the S2 layer to model the neural response of V4 area. Furthermore, some

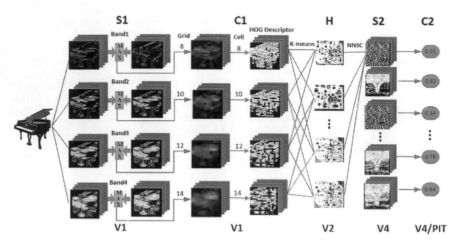

FIGURE 4.3   Structure of the proposed model.

modifications are made in C1 layer and template learning stage. The architecture of our proposed model is described in Fig. 4.3.

In S1 layer, the input image is convolved with Gabor filters at 8 scales and 4 orientations. In C1 layer, the image is first subsampled through a local max operation, and then image patches are described by HOG descriptor. A new layer (H layer) is added between C1 and S2 layers using K-means clustering algorithm. In S2 layer, the unit responses are computed by non-negative sparse coding using image templates as the basis. In C2 layer, C2 unit responses are calculated by pooling the coefficient matrix obtained in S2 layer.

4.3.1.1 Modelling of V1 neurons

As in the HMAX model[12], we use a bank of Gabor filters to emulate the properties of V1 neurons. Different from the HMAX model, we make two modifications in the first two layers. First, we use fewer Gabor filters and only use the first 4 scale bands to improve time efficiency in S1 layer. Second, each C1 image patch is a HOG descriptor[38] rather than natural images in C1 layer.

4.3.1.2 Modelling of V2 neurons

To reproduce the properties of V2 neurons, we add a hidden layer between C1 and S2 layers, denoted as H layer. We use multiple firing K-means to describe the response properties of V2 neurons. Different from

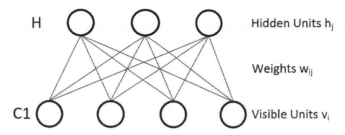

FIGURE 4.4 Illustration of the multiple firing K-means model.

the original K-means, multiple firing K-means allows multiple hidden units fire together for an input[9]. The structure of the multiple firing K-means model is illustrated in Fig. 3.4. The C1 units from the previous layer are input as visible units. For each input unit, there are L hidden units firing. For each input unit, v_i (Fig. 4.4):

$$h_j(v_i) = \begin{cases} 1 & v_i \in h_j \\ 0 & else \end{cases} \qquad (4.5)$$

4.3.1.3 Modelling of V4 neurons

We apply non-negative sparse coding in the S2 layer for calculating the S2 unit responses. For each afferent image patch, we seek a non-negative sparse representation by using the image templates extracted in the learning stage as the basis matrix. Thus the corresponding coefficients are calculated as the unit response between an image patch and a template.

Assume the M afferent image patches $X = [x_1, x_2, \cdots, x_M] (x_i \in R^{D \times 1})$ are the set of vectors of D dimensions. We write all the N templates in the matrix form:

$$T = [t_1, t_2, \cdots, t_N] = \begin{bmatrix} t_{11} & t_{12} & \cdots & t_{1N} \\ t_{21} & t_{22} & \cdots & t_{2N} \\ \vdots & \vdots & \vdots & \vdots \\ t_{D1} & t_{D2} & \cdots & t_{DN} \end{bmatrix} \qquad (4.6)$$

Then the image patches X can be linearly represented by T:

$$X = TC = \begin{bmatrix} t_{11} & t_{12} & \cdots & t_{1N} \\ t_{21} & t_{22} & \cdots & t_{2N} \\ \vdots & \vdots & \vdots & \vdots \\ t_{D1} & t_{D2} & \cdots & t_{DN} \end{bmatrix} \cdot \begin{bmatrix} c_{11} & c_{12} & \cdots & c_{1M} \\ c_{21} & c_{22} & \cdots & c_{2M} \\ \vdots & \vdots & \vdots & \vdots \\ c_{N1} & c_{N2} & \cdots & c_{NM} \end{bmatrix} \qquad (4.7)$$

where C is the non-negative sparse coefficients which can be obtained by solving the following optimization problem[39]:

$$\text{arg. min } \|X - TC\|^2 + \lambda \|C\|_1 \qquad (4.8)$$

Here, we follow the algorithm in[30] to solve this optimization problem.

In C2 layer, we calculate the C2 unit responses by pooling the coefficient matrix C with row-wise maximum:

$$R = \max_j C = (\max_j c_{1j}, \max_j c_{2j}, \cdots, \max_j c_{Nj}) \qquad (4.9)$$

3.3.2 Template selection method

In the original HMAX model, image templates are selected randomly from the training images at different locations and scales[40]. Feature selection plays important roles in object classification[41]. To select more informative templates, we proposed an effective method based on NNSC.

First, all available image patches are extracted from a set of training images. Next, we represent these image patches with non-negative sparse coding. Assume the M available image patches $X = [x_1, x_2, \cdots, x_M] (x_i \in R^{D \times 1})$ are the set of vectors of D dimensions. We can find a basis W ($W \in R^{D \times K}$) and a coefficient matrix S ($S \in R^{K \times M}$) following the method in Mairal et al[35].

Thus the available image patches X can be linearly represented by W:

$$X = WS = \begin{bmatrix} w_{11} & w_{12} & \cdots & w_{1K} \\ w_{21} & w_{22} & \cdots & w_{2K} \\ \vdots & \vdots & \vdots & \vdots \\ w_{D1} & w_{D2} & \cdots & w_{DK} \end{bmatrix} \cdot \begin{bmatrix} s_{11} & s_{12} & \cdots & s_{1M} \\ s_{21} & s_{22} & \cdots & s_{2M} \\ \vdots & \vdots & \vdots & \vdots \\ s_{K1} & s_{K2} & \cdots & s_{KM} \end{bmatrix} \qquad (4.10)$$

Then we select N vectors from W with the largest coefficient as the templates, which can be achieved by pooling the coefficient matrix S with row-wise maximum. By the above steps, we can select the templates which are more informative than those selected randomly.

4.4 RESULTS AND DISCUSSION

We conduct experiments on three benchmark databases: Caltech101[42], Caltech256[43], and GRAZ-01[13]. We compare our proposed model with several prevalent methods, including HMAX[12], EBIM[13], MAX-Pooling[26], VisNet[44], and FBTP-NN model[45]. The proposed model is employed as the feature extraction process and linear SVM[46] is used as the classifier. Visual features are fed to the linear SVM for classification[46-48]. All experiments are run five times and the results are averaged. Parameters of our model consist of four scale bands (from 7×7 to 21×21) and four orientations ($0°$, $45°$, $90°$, $135°$) (for Gabor filters. For HOG algorithm, each block contains 2×2 cells and each cell has 9 bins, which will result in a 36-dimensional vector. The cell size is the same as the spatial pooling grid of corresponding scale band. In the NNSC algorithm, the regularization parameter λ is an important parameter to be tuned. We choose $\lambda = 1.2/\sqrt{m}$ in all experiments and m denotes the size of image patches. $1/\sqrt{m}$ is a classical normalization factor and the constant 1.2 has proved to generate reasonable sparsities[35] (about 10 nonzero coefficients). In our method, the size of image patch is 36, so the regularization parameter $\lambda = 0.2$. The number of image templates is detailed in each experiment.

4.4.1 Caltech101 database

Caltech101 database contains images of 101 object categories and a background category. Each object category consists of 40–800 images and each image is about 300×200 pixels[42]. For all images, the height is resized to 140 and the width is resized accordingly.

4.4.1.1 Present/absent classification

To evaluate the role of proposed methods (K-means, NNSC, and HOG) in our architecture on the classification performance and tuning parameters for K-means algorithm, the proposed methods are tested at different layers in this experiment. As described above, K-means is applied in the H layer modelling V2 neural response; NNSC is used in S2 layer to model the V2 neurons and HOG descriptor is integrated into C1 layer to characterize higher-level features. We select six image categories (see examples in Fig. 4.5) from Caltech101 database as positive sets and background as negative set. Twenty images are selected randomly from each category for training and 30 images for testing.

FIGURE 4.5   Example images from caltech101 database: grand piano, pyramid, nautilus, sunflower, French horn, and pizza.

First we test the role of different methods in the proposed model. We use a single method or combination of the methods including multiple firing K-means, HOG, and NNSC. Fig. 4.6 shows the recognition results of the original HMAX model and our proposed methods with varying number of templates (5, 10, 50, 100, 150, and 200). When using NNSC in S2 layer alone, the recognition performance is relatively poor and is not as good as that of the HMAX model in general. When combing HOG with NNSC, denoted as NNSC + HOG, the recognition performance has greatly improved and significantly outperforms the HMAX model with more than 50 templates. By combing the three methods together denoted as NNSC + K-means + HOG, further, improvement can be obtained. But when there are small numbers of templates (5 or 10), our methods do not have many advantages over the HMAX model.

In our model, the image templates are used as a basis matrix for computing the S2 unit responses. Thus, each image template corresponds to a basis vector. For sparse coding, the number of basis vectors N should be larger than its dimensionality D^{33}. As mentioned above, the

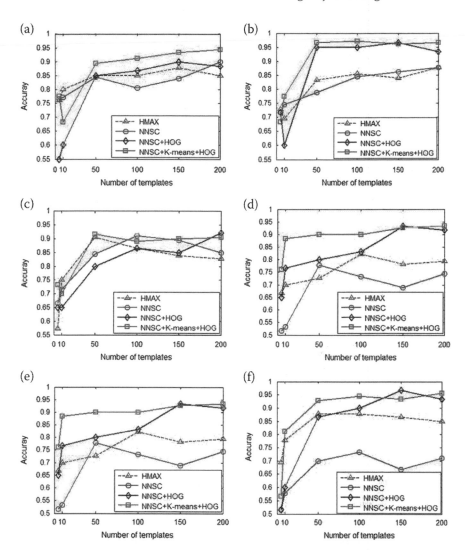

FIGURE 4.6   Recognition results of the original HMAX model and our proposed methods on Caltech101 database. (a) Grand piano; (b) pyramid; (c) nautilus; (d) sunflower; (e) French horn; (f) pizza.

dimensionality of the basis vector is 36. When the number of basis vectors (image templates) is smaller than its dimensionality D, for example, 5 or 10, the image information cannot be characterized enough. So our method does not perform well with small number of templates.

Figure 4.7 presents the time efficiency comparison of the original HMAX model and our proposed methods on the pyramid category. As

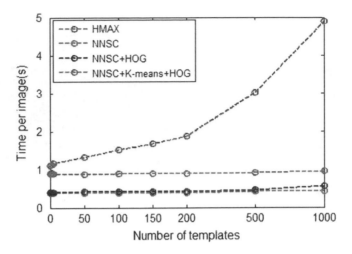

FIGURE 4.7 Time efficiency comparison of the original HMAX model and our proposed methods on pyramid category.

the number of templates increases, the processing time of our methods increases slightly and almost remains the same. But to the HMAX model, the processing time increases obviously with the number of templates. When the number of templates is small, the improvement of processing speed is not very obvious. When the number of image templates is large, for example, 1000, the speed is greatly enhanced. Especially, for the combination of three methods (NNSC + K-means + HOG), the speed is about 5 times of the HMAX model.

Fig. 4.8 shows the comparison of processing speed between NNSC and Euclidean distance in computing S2 unit response. Test results show that NNSC is much more efficient than Euclidean distance in all conditions. When there are a large number of templates or image patches, the improvement of processing speed is more obvious. The processing speed of our method is at least 15 times faster than that of Euclidean distance under the same condition.

So our proposed method has advantage in both recognition performance and time efficiency especially in case of large number of templates. The improvement of time efficiency makes our model applicable to practical applications. For convenience, we denote our proposed model, the combination of NNSC, K-means, and HOG as Sparse-HMAX in the following experiment.

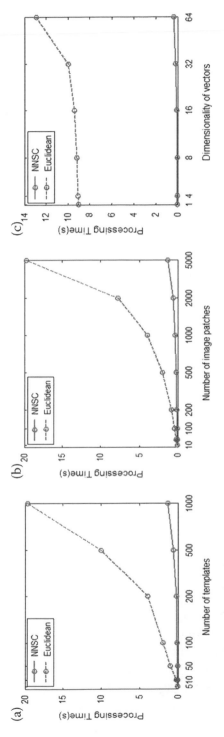

FIGURE 4.8 Comparison of processing speed between NNSC and Euclidean distance in computing S2 unit response. (a) Various number of templates; (b) various number of image patches; (c) different dimensionality of vector.

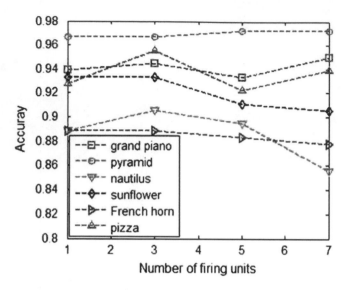

FIGURE 4.9 Recognition performance of six object categories with various number of firing units.

Fig. 4.9 presents the recognition performance of our method with different number of firing units in H layer. The recognition performance does not make a significant difference among various number of firing units with a fixed number of hidden units. Relatively speaking, we can get a better recognition performance with 3 firing units for each input unit. Larger number of firing units will take more computational time. So in consideration of recognition performance and time efficiency, 3 hidden firing units are used for each input in the H layer in the following experiments.

4.4.1.2 Binary classification

We conduct experiment on the Caltech101 database (see Fie-Fie et al.[42] for details) compared with two bio-inspired models: HMAX model and Max-Pooling model[26]. As described in[26], four pairs of image categories are selected for this experiment, including airplane and motorbike, chair and grand piano, elephant and crocodile, bass and dolphin (see examples in Fig. 4.10). For the airplane/motorbike pair, there are 200 images chosen from each category. For the other three pairs, 50 images are selected from each category. Half are selected randomly for training from each category and the rest for testing.

FIGURE 4.10   Example images of four pairs of categories from Caltech101 database. From left to right are respectively airplane/motorbike, chair/grand piano, elephant/crocodile, and bass/ dolphin.

In our method, the templates used in S2 layer are selected using NNSC based learning method. The feature vectors are formed as same as[12] and the number of templates varies from 10 to 200. The Max-Pooling method is as described in Tang and Qiao[26].

Table 4.1 presents the performance comparison of Sparse-HMAX, HMAX, and Max-Pooling with 200 image templates and Fig. 4.11 demonstrates the performance comparison of the three models with varying number of templates. Results show that our method significantly outperforms the HMAX for all pairs with more than 50 templates and achieves better recognition performance than the Max-Pooling model.

4.4.1.3 Multiclass classification

To get an overall classification performance, we choose 15 images randomly per category for training and the rest of the images are used for testing. The results are reported on 1000 randomly selected features and 5 independent trials. The classification result for 15 training images per category is 50%. This greatly outperforms the HMAX model whose classification results are 35%[49].

TABLE 4.1   Performance comparison of four pair categories on Caltech101 database

Method	Average accuracy of 5 tests (%)			
	Airplane/motorbike	Chair/grand piano	Crocodile/elephant	Bass/dolphin
HMAX	93.7	86.0	84.0	79.0
Max-Pooling	98.3	97.6	93.2	81.2
Sparse-HMAX	**99.0**	**98**	**94.5**	**85.6**

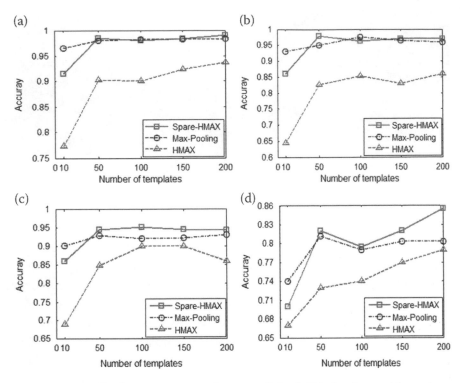

FIGURE 4.11 Performance comparison on Caltech101 database with various number of templates: (a) Airplane/motorbike; (b) chair/grand piano; (c) crocodile/elephant; (d) bass/dolphin.

4.4.2 Caltech256 database

The Caltech256 database consists of 256 object categories and brings great challenges to object recognition[43]. The images have great variation in size, position, illumination, poses, and viewpoint.

4.4.2.1 Binary classification

The proposed method together with a full HMAX model, a scaled-down HMAX model with a comparable complexity to that of Sparse-HMAX (denoted as HMAX_min) and VisNet model[44] are used to perform classification tasks between two object categories from Caltech256 database, cowboy-hat and teddy-bear (see examples in Fig. 4.12). We follow the same way in Rolls[44] and the parameters of three compared models are also the same as in Rolls[44]. We use 200 image templates in our model. 60 images are chosen randomly from each category and resized to

FIGURE 4.12 Example images of teddy-bear and cowboy hat.

256×256. The selected images are grouped into training and testing samples, with varying number (1, 5, 15, and 30) of training samples, and the remaining images are used for testing. In our model, there are approximately 22,600 units in the C1 layer, 3200 units in the S2 layer, and 200 units in the C2 layer, which is less than half of the 65,536 neurons of VisNet.

Fig. 4.13 presents the classification performance of our model and compared models on the Caltech256 database. The sparse-HMAX model greatly outperforms HMAX_min with more than 5 training images.

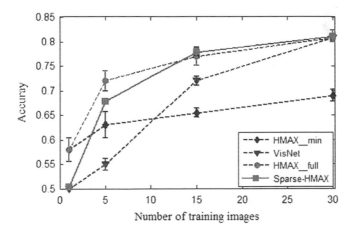

FIGURE 4.13 Classification performance of Sparse-HMAX, VisNet, full HMAX, and HMAX_min on the Caltech256 database.

Our model also exhibits higher performance than VisNet model with small number of units. The full HMAX model (given its very large number of units and large number of templates) shows better performance than that of Sparse-HMAX with small number of training images. But once there is reasonable number of training images (more than 5), our method shows equivalent performance with the full HMAX model.

4.4.2.2 Three-class classification

To evaluate the classification performance of Sparse-HMAX model under more complex conditions, we conduct a three-class classification experiment on Caltech256 database. As described in[45], we select three object categories (bicycle, treadmill, and revolver) from Caltech256 database, as shown in Fig. 4.14. For each category, 50 object images with various sizes, orientations, appearances, and backgrounds are selected. 50% of the images are for training and the remaining 50% for testing. All images keep their original size. Table 4.2 shows the recognition results of the proposed method compared with FBTP-NN[45] and HMAX model.

FIGURE 4.14 Example images taken from caltech256 database. From top to bottom, the categories are respectively bicycle, treadmill, and revolver.

TABLE 4.2   Recognition results for three-class experiments

Methods	Accuracy(%)
HMAX[12]	77.33
FBTP-NN	92.31
Sparse-HMAX	94.13

Experimental result shows that our proposed model has significantly improvement on the HMAX model. In contrast with the FBTP-NN, our method also shows better recognition performance.

4.4.3 GRAZ-01 database

GRAZ-01 is a challenging dataset[50] and contains three categories: bikes, persons, and backgrounds[51]. Some sample images are shown in Fig. 4.15. We follow the same strategy used in Huang et al.[13]. A hundred positive images (bikes or persons) and 100 negative images (backgrounds) are selected randomly for training. Fifty other positive images and negative images are chosen for testing. A total of 1500 image templates are used in this experiment. All images keep their original size. We compare our method with two related methods (the standard HMAX model and EBIM[13]).

Experimental results are presented in Table 4.3 and Fig. 4.16. Table 4.3 shows the comparison among several bio-inspired methods in

FIGURE 4.15   Example images taken from GRAZ-01. The categories are bikes, people, and backgrounds from top to bottom.

TABLE 4.3   Performance comparison of several methods on GRAZ-01 database. The second row of performance is reported in Huang et al.[13]

Method	Bikes		Persons	
	EER	AUC	EER	AUC
HMAX[12]	82.8	89.6	85.0	91.2
EBIM[13]	84.1	90.5	86.0	91.8
Sparse-HMAX	**86.8**	**93.1**	**86.6**	**93.2**

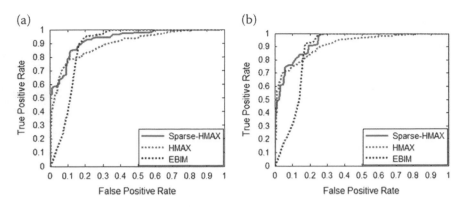

FIGURE 4.16   ROC curves of Sparse-HMAX model and compared models on GRAZ-01. (a) Bikes, (b) persons.

EER (detection rate at equal-error-rate of the ROC curve) and AUC (area under the ROC curve). Results show the Sparse-HMAX outperforms HMAX in all cases. Table 4.3 and Fig. 4.15 demonstrate that Sparse-HMAX outperforms the EBIM model for persons. The ROC curves of Sparse-HMAX are comparable to that of EBIM for bikes. Generally, Sparse-HMAX model achieves comparable results compared with EBIM model.

4.4.4 Template selection method

We evaluate the effectiveness of template selection method compared with other methods including k-means[40] and random selection method[12]. We perform experiments on four object categories (see examples in Fig. 4.17) of the challenging Caltech256 database. We use the four object categories as positive sets and background as negative set. For each category, 20 images are selected randomly for training and 30 images for testing.

Fig. 4.18 presents the comparison results of our method with k-means and random selection method. Our method outperforms the K-means and random selection method for all object categories.

4.5 CONCLUSIONS

In this chapter, we propose a biologically inspired model, denoted as Sparse-HMAX, to imitate the process of object recognition in visual cortex. We extend the standard HMAX model in some biologically

FIGURE 4.17 Example images from Caltech256. From top to bottom, the object categories are bat, bear, cake, and chopstick.

feasible ways. There are many biologically inspired models capable of producing the receptive fields of V1 and V4 neurons in visual cortex[12,13,25,25,25,50], but few concerns with modelling the receptive fields of V2 neurons. By using multiple firing K-means between C1 and S2 layer, our model has the capability of reproducing the response properties of V2 neuron and is more biologically feasible than other computer vision models[25,43,44]. By integrating NNSC into S2 layer, our model can better interpret the response properties of V4 neurons than the original HMAX. More local features can be extracted with HOG-based image patches and more informative templates can be learned during the learning stage with NNSC based selection method.

We take some measures to improve the time efficiency of our model. In S1 layer, we use less Gabor filter bands and only use half as many scale bands as HMAX. Thus, our method doubles the processing efficiency on the whole. In S2 layer, we use a different strategy to computer S2 unit response by using NNSC instead of Euclidean distance. Our method is much more efficient than Euclidean distance in computing S2 response. Test results show that the processing speed of our method is at least 15 times faster than that of Euclidean distance under the same condition. Though H layer takes some extra time in our model, it reduces the number of C1 image patches at the same time.

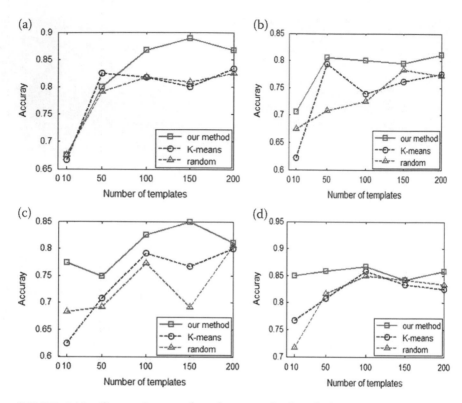

FIGURE 4.18 Comparion results of our method with k-means and random selection method on Caltech256 database. (a) Bat; (b) bear; (c) cake; (d) chopstick.

To evaluate the performance of our method, we conduct experiments on three public databases. Experimental results on Caltech101 database show that our proposed method not only significantly outperforms the original HMAX model on both recognition performance and time efficiency, but also has advantage on Max-Pooling model[26]. Experimental results on Caltech256 database demonstrate that our model shows better recognition performance than HMAX, VisNet[44], and FBTP-NN[45] model. Experimental results on the challenging GRAZ-01 database show that our model achieves comparable results compared with EBIM model. It is worth noting that our method significantly outperforms the original HMAX model in all object recognition tasks. Although our model does not exhibit the best in all tasks, it outperforms some state-of-art methods in most cases.

In the future, we would like to improve the Sparse-HMAX as follows. Due to the strict constraint of k-means algorithm, we will relax the constraint for further improvement perhaps by using sparse coding instead of multiple firing K-means in modelling V2 neurons. Moreover, considering the important role of the templates to the recognition performance, we will search for more effective feature learning strategies.

REFERENCES

1. Yu, J., Rui, Y., Tang, Y. Y., and Tao, D. 2014. High-order distance based multiview stochastic learning in image classification. *IEEE Transactions on Cybernetics* 44(12):2431–2442.
2. Yamins, D., Hong, H., Cadieu, C. F., Solomon, E. A., Seibert, D., and Dicarlo, J. J. 2014. Performance-optimized hierarchical models predict neural responses in higher visual cortex. *Proceedings of the National Academy of Sciences* 111(23):8619–8624.
3. Rolls, E.T. 2000. Functions of the primate temporal lobe cortical visual areas in invariant visual object and face recognition, *Neuron* 27(2):205–218.
4. Ito, M., and Komatsu, H. 2004. Representation of angles embedded within contour stimuli in area V2 of macaque monkeys. *The Journal of Neuroscience* 24(13):3313–3324.
5. Pasupathy, A., and Connor, C. E. 2002. Population coding of shape in area V4. *Nature Neuroscience* 5(12):1332–1338.
6. Lau, K. H., Yong, H. T., and Lo, F. L. 2014. Sparsity-regularized HMAX for visual recognition, *PLoS ONE* 9(1).
7. Rust, N. C., and Di Ca Rlo, J. J. 2010. Selectivity and tolerance ("invariance") both increase as visual information propagates from cortical area v4 to it. *The Journal of Neuroscience: The Official Journal of the Society for Neuroscience* 30(39):12978–12995.
8. Tao, D., Li, X., Wu, X., and Maybank, S.J. 2007. General tensor discriminant analysis and Gabor features for gait recognition, *IEEE Transactions on Pattern Analysis and Machine Intelligence* 29(10):1700–1715.
9. Hu, X., Zhang, J., Peng, Q., and Bo, Z. 2014. Modelling response properties of v2 neurons using a hierarchical k-means model. *Neurocomputing* 134(Jun.25):198–205.
10. Riesenhuber, M., and Poggio, T. 1999. Hierarchical models of object recognition in cortex, *Nat. Neurosci.* 2(11):1019–1025.
11. Cadieu, C., Kouh, M., Pasupathy, A., Connor, C. E., Riesenhuber, M., and Poggio, T. 2007. A model of V4 shape selectivity and invariance. *Journal of Neurophysiology* 98(3):1733–1750.

12. Serre, T., Wolf, L., Bileschi, S., Riesenhuber, M., and Poggio, T. 2007. Robust object recognition with cortex-like mechanisms. *IEEE Transactions on Pattern Analysis and Machine Intelligence* 29(3):411–426.
13. Huang, Y., Huang, K., Cheng, D., Tan, T., and Li, X. 2011. Enhanced biologically inspired model for object recognition, *IEEE Transactions on Pattern Analysis and Machine Intelligence* 41(6):1668–1680.
14. Freeman, J., and Simoncelli, E. P. 2011. Metamers of the ventral stream. *Nature Neuroscience* 14(9):1195–1201.
15. Hinton, G. E., Osindero, S., and Teh, Y. W. 2014. A fast learning algorithm for deep belief nets. *Neural Computation* 18(7):1527–1554.
16. Ekanadham, C. 2008. Sparse deep belief net models for visual area v2. *Advances in Neural Information Processing Systems* 20.
17. Hinton, G.E. 2002. Training products of experts by minimizing contrastive divergence. *Neural Computation* 14 (8):1771–1800.
18. Mutch, J., and Lowe, D.G. 2008. Object class recognition and localization using sparse features with limited receptive fields. *International Journal of Computer Vision* 80(1): 45–57.
19. Zhang, C., Liu, J., Tian, Q., Xu, C., Lu, H., and Ma, S. 2011, June. Image classification by non-negative sparse coding, low-rank and sparse decomposition. *In CVPR* 2011:1673–1680.
20. Olshausen, B. A., and Field, D. J. 1996. Emergence of simple-cell receptive field properties by learning a sparse code for natural images. *Nature* 381(6583):607–609.
21. Olshausen, B. A., and Field, D. J. 1997. Sparse coding with an overcomplete basis set: a strategy employed by V1? *Vision Research* 37(23): 3311–3325.
22. Carlson, E. T., Rasquinha, R. J., Zhang, K., and Connor, C. E. 2011. A sparse object coding scheme in area V4. *Current Biology* 21(4):288–293.
23. Yang, J., Kai, Y., Gong, Y., and Huang, T. S. 2009. Linear spatial pyramid matching using sparse coding for image classification. 2009 IEEE Computer Society Conference on Computer Vision and Pattern Recognition (CVPR 2009):20–25.
24. Wang, Y., and Deng, L. 2016. Modelling object recognition in visual cortex using multiple firing K-means and non-negative sparse coding. *Signal Processing* 124:198–209.
25. Ghodrati, M., Khaligh-Razavi, S. M., Ebrahimpour, R., Rajaei, K., and Pooyan, M. 2012. How can selection of biologically inspired features improve the performance of a robust object recognition model? *PLoS ONE* 7(2):e32357.
26. Tang, T., and Qiao, H. 2014. Improving invariance in visual classification with biologically inspired mechanism. *Neurocomputing* 133:328–341.
27. Shastri, B. J., and Levine, M. D. 2007. Face recognition using localized features based on non-negative sparse coding. *Machine Vision and Applications* 18(2):107–122.

28. Hoyer, P. O. 2002, September. Non-negative sparse coding. In Proceedings of the 12th IEEE Workshop on Neural Networks for Signal Processing: 557–565.
29. Guan, N., Tao. D., Luo, Z., and Yuan, B. 2012. NeNMF: an optimal gradient method for nonnegative matrix factorization. *IEEE Transactions on Signal Processing* 60(6):2882–2898.
30. Olshausen, B. A., and Field, D. J. 1997. Sparse coding with an over-complete basis set: A strategy employed by V1? *Vision Research* 37(23):3311–3325.
31. Gregor, K., and LeCun, Y. 2010, June. Learning fast approximations of sparse coding. In Proceedings of the 27th international conference on international conference on machine learning: 399–406.
32. Song, X., Liu, Z., Yang, X., et al. 2014. A new sparse representation-based classification algorithm using iterative class elimination. *Neural Computing and Applications* 24(7-8):1627–1637.
33. Yang, J., Peng, Y., Xu, W., and Dai, Q., 2009. Ways to sparse representation: an overview. *Science in China Series F: Information Sciences* 52(4): 695–703.
34. Yu, J., Rui, Y., and Tao, D. 2014. Click prediction for web image re-ranking using multimodal sparse coding. *IEEE Transactions on Image Processing* 23(5): 2019–2032.
35. Mairal, J., Bach, F., Ponce, J., and Sapiro, G. 2010. Online learning for matrix factorization and sparse coding. *Journal of Machine Learning Research* 11(1):19–60.
36. Guan, N., Tao, D., Luo, Z., and Yuan, B. 2011. Manifold regularized discriminative nonnegative matrix factorization with fast gradient descent. *IEEE Transactions on Image Processing* 20(7):2030–2048.
37. Guan, N., Tao, D., Luo, Z., and Yuan, B. 2012. Online nonnegative matrix factorization with robust stochastic approximation. *IEEE Transactions on Neural Networks and Learning Systems* 23(7):1087–1099.
38. Dalal, N., and Triggs, B. 2005, June. Histograms of oriented gradients for human detection. In 2005 IEEE computer society conference on computer vision and pattern recognition (CVPR'05), 1:886–893.
39. Li, H., Li, H., Wei, Y., Tang, Y., and Wang, Q. 2014. Sparse-based neural response for image classification. *Neurocomputing* 144: 198–207.
40. Mishra, P., and Jenkins, B. K. 2010, March. Hierarchical model for object recognition based on natural-stimuli adapted filters. In 2010 IEEE International Conference on Acoustics, Speech and Signal Processing: 950–953.
41. Tao, D., Li, X., Wu, X., and Maybank, S. J. 2008. Geometric mean for subspace selection. *IEEE Transactions on Pattern Analysis and Machine Intelligence* 31(2):260–274.
42. Fei-Fei, L., Fergus, R., and Perona, P. 2004, June. Learning generative visual models from few training examples: An incremental bayesian

approach tested on 101 object categories. In 2004 conference on computer vision and pattern recognition workshop, 178.

43. Griffin, G., Holub, A., and Perona, P. 2007. Caltech-256 object category dataset. Caltech Technical Report, Los Angeles: 1–20.
44. Rolls, E. T. 2012. Invariant visual object and face recognition: neural and computational bases, and a model, VisNet. *Frontiers in Computational Neuroscience* 6:35.
45. Zheng, Y., Meng, Y., and Jin, Y. 2011. Object recognition using a bio-inspired neuron model with bottom-up and top-down pathways. *Neurocomputing* 74(17):3158–3169.
46. Tao, D., Tang, X., Li, X., and Wu, X. 2006. Asymmetric bagging and random subspace for support vector machines-based relevance feedback in image retrieval. *IEEE Transactions on Pattern Analysis and Machine Intelligence* 28(7): 1088–1099.
47. Tao, D., Jin, L., Liu, W., and Li, X. 2013. Hessian regularized support vector machines for mobile image annotation on the cloud. *IEEE Transactions on Multimedia* 15(4):833–844.
48. Yu, J., Tao, D., Wang, M., and Rui, Y. 2014. Learning to rank using user clicks and visual features for image retrieval. *IEEE Transactions on Cybernetics* 5(4):67–779.
49. Theriault, C., Thome, N., and Cord, M. 2012. Extended coding and pooling in the HMAX model. *IEEE Transactions on Image Processing* 2(2):64–777.
50. Opelt, A., Pinz, A., Fussenegger, M., Auer, P. 2006. Generic object recognition with boosting. *IEEE Transactions on Pattern Analysis and Machine Intelligence* 8(3): 6–431.
51. Lu, Y. F., Zhang, H. Z., Kang, T. K., Choi, I. H., and Lim, M. T. 2014. Extended biologically inspired model for object recognition based on oriented Gaussian–Hermite moment. *Neurocomputing* 39:89–201.

APPENDIX

```
%% Programmed by Matlab
```

1. demoRelease.m
```
%demonstrates how to use sparse-HMAX model features
in a pattern classification framework
addpath ('osu-svm') %put your own path to osusvm here
addpath('HOG');
useSVM = 1; %if you do not have osusvm installed you
can turn this
```

```
          %to 0, so that the classifier would be a NN
classifier
  %note: NN is not a great classifier for these features
READPATCHESFROMFILE = 0; %use patches that were al-
ready computed
                %(e.g., from natural images)
patchSizes = [12]; %all sizes are required
numPatchSizes = length(patchSizes);
numPatchesPerSize = 200;
%specify directories for training and testing images
train_set = 'E:\testImage\caltech256-3\train'; %
path for training
test_set = 'E:\testImage\caltech256-3\test'; %  path
for testing
%patch_set = 'E:\testImage\patch'; %path for patch
[dnames,len] = GetDir(train_set,test_set); %cI is a
cell containing all training and testing images
pI = HreadPatchImages(dnames{1});
if isempty(pI{1})
 error(['No training images were loaded -- did you
remember to' ...
' change the path names?']);
end
%----Settings for Testing --------%
rot = [90 -45 0 45];
%rot = [0];
c1ScaleSS = [1:2:9];
RF_siz    = [7:2:23];
c1SpaceSS = [8:2:14];
minFS     = 7;
maxFS     = 23;
div = [4:-.05:3.6];
Div     = div;
fprintf(1,'Initializing gabor filters --
full set...');
%creates the gabor filters use to extract the S1 layer
[fSiz,filters,c1OL,numSimpleFilters]   = init_gabor
(rot, RF_siz, Div);
```

```
fprintf(1,'done\n');
%-extract prototype
cPatches = extractPrototype5(pI{1}, numPatchSizes,
numPatchesPerSize, patchSizes);
%The actual C2 features are computed below for each
one of the training/testing directories
for i = 1:2*len,
 pI=HreadPatchImages(dnames{i});
   C2res{i} = extractC2forcell51(cPatches,pI{1},
numPatchSizes,numPatchesPerSize, patchSizes);
end
clear pI cPatches
%Multi-classification code
XTrain=[];
XTest=[];
TrainLabel=[];
TestLabel=[];
for i=1:len
 XTrain=[XTrain C2res{i}];
 XTest=[XTest C2res{i+len}];
   TrainLabel   =   [TrainLabel;i*ones(size(C2res
{i},2),1)];%testing examples as columns
 TestLabel=[TestLabel;i*ones(size(C2res
{i+len},2),1)];
end
clear C2res
%% libsvm classification
XTrain=XTrain';
XTest=XTest';
model = svmtrain(TrainLabel,XTrain,'-s 0 -t 2 -c 1.2
-g 2.8');
[predictlabel,accuracy, dec_values] = svmpredict
(TestLabel,XTest,model);
```

2. extractPrototype5
```
function cPatches = extractPrototype5
(cItrainingOnly, numPatchSizes, numPatchesPerSize,
patchSizes);
```

```
%%Extract prototype using NNSC as part of the
training of the C2 classification system.
%Note: we extract only from BAND 2. Extracting from
all bands might help
%cPatches the returned prototypes
%cItrainingOnly the training images
%numPatchesSizes is the number of patches sizes
%numPatchesPerSize is the number of prototypes ex-
tracted for each size
%patchSizes is the vector of the patches sizes
if nargin<2
 numPatchSizes - 4;
 numPatchesPerSize = 250;
 patchSizes = 4:4:16;
end
nImages = length(cItrainingOnly);
%----Settings      for     Training     the    random
patches--------%
rot = [90 -45 0 45];
c1ScaleSS = [1 3];
RF_siz   = [11 13];
c1SpaceSS = [10];
minFS    = 11;
maxFS    = 13;
div = [4:-.05:3.2];
Div    = div(3:4);
%fprintf(1,'Initializing  gabor  filters  --  par-
tial set...');
[fSiz,filters,c1OL,numSimpleFilters]  =  init_gabor
(rot, RF_siz, Div);
%fprintf(1,'done\n');
%cPatches = cell(numPatchSizes,1);
cPatches = cell(numPatchSizes,numPatchesPerSize);
bsize = [0 0];
band=length(c1SpaceSS);
% c1Ims=cell(1,4*nImages);% ipts=[];
for ii=1:nImages,% for each training image
  stim = cItrainingOnly{ii};
```

```
    [c1Image,c1source,s1source] = C1(stim, filters,
fSiz, c1SpaceSS, c1ScaleSS, c1OL);
        %c1source   =   C51(stim,   filters,   fSiz,
c1SpaceSS,c1ScaleSS, c1OL);
  for i=1:length(c1Image)%1:32
    n=(ii-1)*length(c1Image);
    c1Ims{1,i+n}=c1Image{1,i};
    c1Ims{1,i+n}=im2double(c1Ims{1,i+n});
  end
end
%save c1Ims.mat c1Ims;
cPatches=Sparse_coding_new6(c1Ims,numPatches
PerSize,0.2,patchSizes);
```

3. extractC2forcell5.m

```
function mC2 = extractC2forcell5(cPrototype,
cImages,numPatchSizes,numPatchesPerSize,
patchSizes);
%
%this function is a wrapper of C2. For each image in
the cell cImages,
%it extracts all the values of the C2 layer
%for all the prototypes in the cell cPatches.
%The result mC2 is a matrix of size total_number_-
of_patches \times number_of_images where
%total_number_of_patches  is  the  sum  over i  =
1:numPatchSizes of length(cPatches{i})
%and number_of_images is length(cImages)
%The C1 parameters used are given as the variables
filters,fSiz,c1SpaceSS,c1ScaleSS,c1OL
%for more detail regarding these parameters see the
help entry for C1
%
%See also C1
%{
numPatchSizes = min(numPatchSizes,length
(cPatches));%length() is the row
```

```
%all the patches are being flipped. This is becuase in
matlab conv2 is much faster than filter2
for i = 1:numPatchSizes,%numPatchSizes=4
 [siz,numpatch] = size(cPatches{i});%[4,250]
 siz = sqrt(siz/4);%siz=1
 for j = 1:numpatch,
   tmp = reshape(cPatches{i}(:,j),[siz,siz,4]);
   tmp = tmp(end:-1:1,end:-1:1,:);
   cPatches{i}(:,j) = tmp(:);
 end
end
%}
mC2 = [];
for i = 1:length(cImages), %for every input image
 fprintf(1,'%d:',i);
 stim{1} = cImages{i};
 img_siz = size(stim);
 iC2 = []; %bug fix

 c1Patch = extractPrototype6(stim, numPatchSizes,
numPatchesPerSize, patchSizes,cPrototype);
 mC2 = [mC2, c1Patch];
end
```

4. Sparse_coding_new6

```
function [I] = Sparse_coding_new6(img_in,nBases,
alpha,psize)
%img_in is the set of input images
%X is input matrix
%B is the dictionary,nBases is the number of
dictionary
%alpha is the sparsity parameter
%maxiter: maximum iteration
%I: the matrix of extracted templates
----------------read input images----------------
imgpath='E: /american-flag/6trainpos/pictures';
pathname=uigetdir(cd,imgpath);
if pathname==0
```

```
msgbox(' You have selected the wrong file folder!!');
  return;
end
  %open image files
filesjpg=ls(strcat(pathname,'\*.jpg'));
files=[cellstr(filesjpg)];
len=length(files);
for i=1:len
  if strcmp(cell2mat(files(i)),'')
    continue;
  end
  filesname{i}=strcat(pathname,'\',files(i));
  img_in{i}=imread(cell2mat(filesname{i}));
end
% --------------------------- Generate the block
matrix of the training image --------------------
len=length(img_in);
numPatchSizes =length(psize);
I=cell(numPatchSizes,nBases);
for s = 1:numPatchSizes
  ipts=[];
  ip=[];
  for i=1:len
    BinNum = 9;
    Angle = 360;
    block=2;
        ip= ImgHOGFeature9(img_in{i}, psize(s)/2,
BinNum, Angle,psize(s),block, [0;0]);
    ipts=[ipts ip];
  end
 X=ipts;
 param.K=500;
 param.lambda=0.3;
 param.iter=30;
 param.modeParam=0;
 K=500;
 [W H] = nnsc(X,param);
 %%
```

```
s1=zeros(K,1);
for i=1:K
  s1(i)=sum(H(i,:));
end
s2=reshape(1:K,1,K);
s2=s2';
s3=[s1 s2];
s3=sortrows(s3,1);

for i=1:nBases
   I{s,i}=W(:,s3(K-i+1,2));
end

end

end
```

Biological modelling of the human visual system using GLoP filters and sparse coding on multi-manifolds

5.1 INTRODUCTION

Humans can perform object recognition effectively and effortlessly no matter how cluttered and complex the conditions are. This remarkable ability is mainly supported by the visual system. But for computers, it is not so easy and involves enormous challenges. The objects to be recognized often vary in position, size, lighting condition, and viewpoint, which pose great difficulty for object recognition. How to mimic the remarkable visual system and extract invariant features representing objects has become the key issue of object recognition[1].

In the recent few years, some new image presentation methods have been proposed for different recognition tasks. Yu et al.[2] extracted scene's

DOI: 10.1201/9781003281641-5

multiview features and explored their complementary characteristics, and further proposed a novel multimodal dimensionality reduction method for natural scene classification. Sang et al.[3] proposed a Ranking based Multi-correlation Tensor Factorization method to describe the trinary relations among image, user, and tag. Sang et al.[4] also proposed a novel visualization scheme to provide a rich landscape for geographical exploration from multiple views. Meanwhile, metric-learning methods have demonstrated powerful performance in real-time systems. Tan et al.[5] proposed a new framework for object recognition based on weakly-supervised metrics and template learning, which demonstrated great advantage of high computational speed and robustness against image noise. In addition, Tan et al.[6] proposed a metric-learning-based template matching method for traffic sign recognition and presented an SVM-based weakly supervised metric-learning method, which achieved encouraging results on multiview traffic sign recognition. Though these methods have achieved outstanding performance in various recognition tasks such as natural scene classification, face recognition, and traffic sign recognition, they have little relation to the neural mechanism of visual cortex and are not biologically inspired.

To mimic the behaviour of human visual system, many computational models have been proposed inspired by neurophysiological and cognitive neuroscience research findings, which include the Neocognitron proposed by Fukushima in 1980[7], an HMAX model which attempted to emulate the mechanism of the visual cortex[8], deep learning neural networks[9,10], and so on.

Recently, Deep Neural Networks (DNNs) have exhibited excellent performance for visual recognition and have been used in many fields, such as object classification[11] and recognition[12], object detection[13], image privacy protection[14], and image ranking[15]. But they are large neural networks with large training datasets and millions of free parameters need to be optimized through an extensive training phase[16]. Relatively speaking, the HMAX model is more lightweight, which only consists of four layers and involves a few parameters to be tuned. Although the HMAX model has exhibited outstanding ability for object recognition and could generate a set of position- and scale-invariant features for later recognition, there are still some limitations. First, the lower level processing such as retina and LGN is not concerned. Second, though the HMAX model can duplicate some neural properties of V1 and V4 areas[17], the properties of higher-level neurons are not specified enough. Moreover, though it is robust to position and scale changes, it is sensitive to rotational change.

Neuroscience has made great achievements in understanding how visual information is processed in the brain. During information processing of the visual system, information representation from the retina via the LGN provides preprocessing for higher-level processing[18]. It proves that human retina and LGN perform important visual information processing such as variation in luminance and contrast across images[19]. We follow the same way to generate more robust information from the input image before feature extraction.

Recent studies suggest that sparse firing is a property of neurons throughout the visual pathway. Sparse coding can not only well interpret the properties of macaque V1 cells[20], but also accounts for the properties of V4 cells[21]. It is also true for neurons in the medial temporal lobe (MTL)[22]. Sparse coding has been used to model neural response in the higher level of visual system and consequently improve recognition performance[23,24]. In addition, sparse coding has achieved promising performance in the challenging action recognition[25] and web image re-ranking[26].

Motivated by cognitive neuroscience research results and preliminary investigations, we extend the HMAX model in the following biologically feasible ways to model the human visual system[9].

1. Given the fact that many LGN neurons are responsive to centre-surround patterns, contrast normalization is performed to simulate the processing of human retina and LGN.

2. GLoP filters are used to simulate the properties of V1 simple cells and increase the robustness to rotation and zooming.

3. Considering the sparse firing properties of V4 neurons and manifold way of visual perception[27], SCMM is used to compute the simple cell response of V4 area instead of Euclidean distance.

4. To select informative templates, a template learning method is proposed based on DLMM.

5.2 METHOD

5.2.1 HMAX model

The standard HMAX is a bio-inspired model with four layers of computational units[8] (see Fig. 5.1), which can be briefly described as follows.

S1 units: S1 units reproduce the simple cells in the primary visual cortex (V1) in form of Gabor functions which are defined as:

$$F(x, y) = \exp\left(-\frac{x_0^2 + y^2 y_0^2}{2\sigma^2}\right)\cos\left(\frac{2\pi}{\lambda}x_0\right),$$

$$x_0 = x\cos\theta + y\sin\theta \text{ and } y_0 = -x\sin\theta + y\cos\theta$$

(5.1)

where λ, θ, σ, and γ are filter parameters which respectively represent the wavelength, orientation, standard deviation, as well as the spatial aspect ratio.

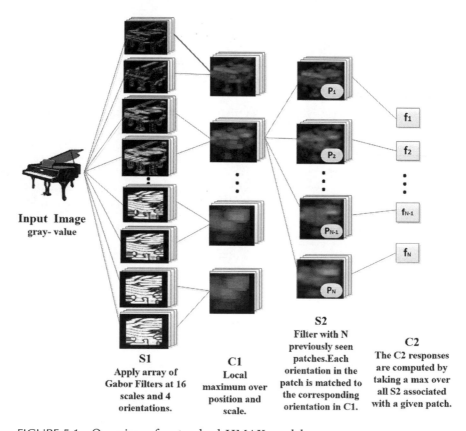

Input Image gray-value

S1	C1	S2	C2
Apply array of Gabor Filters at 16 scales and 4 orientations.	Local maximum over position and scale.	Filter with N previously seen patches. Each orientation in the patch is matched to the corresponding orientation in C1.	The C2 responses are computed by taking a max over all S2 associated with a given patch.

FIGURE 5.1   Overview of a standard HMAX model.

C1 units: C1 units simulate the properties of V1 complex cells which show some tolerance to scale and position change. C1 units are computed by taking a local max-pooling operation[28] over scale and position on the S1 units.

S2 units: S2 unit is computed by the Euclidean distance between an image patch and a stored template, which can be described by:

$$y = \exp(-\beta \| S - P_i \|^2) \tag{5.2}$$

Where β is the tuning sharpness and P_i indicates one of the image templates extracted during the learning stage and S is an image patch at a particular scale from C1 layer.

C2 units: Each C2 unit is obtained by a maximum operation on all S2 units associated with a given template.

5.2.2 The proposed model

5.2.2.1 Problem formulation

The model proposed in this chapter, denoted as E-HMAX, consists of five contributions to the HMAX model. Fig. 5.2 presents the flowchart of the proposed model where the contributions are highlighted in red.

Fig. 5.3 summarizes the overall framework of the E-HMAX model. First, the input image is converted to grey image and processed with contrast normalization using centre-surround operations. Then the normalized image is convolved with a set of GLoP filters with multiple scales and orientations in S1 layer. Here we see convolution at 4 scales and 4 orientations (colour indicates orientation). Next, C1 layer first subsamples the afferent image from the previous layer using a local maximum over adjacent scale and position with the same orientation as in the HMAX model, and then divides the subsampled images into subpatches and extract SIFT descriptor on each patch. In S2 layer, the afferent C1 image patches are presented by SCMM using stored templates as the dictionary, and the learned coefficient matrix is exactly the S2 unit response between C1 image patches and templates. At the last layer, global sum is performed over all positions and scales. For each template, we sum up the unit responses between the given template and all images patches at every position and scale, which will result in a vector of M values (M is the number of image templates). In the following sections, we will give a detailed description of each processing stage.

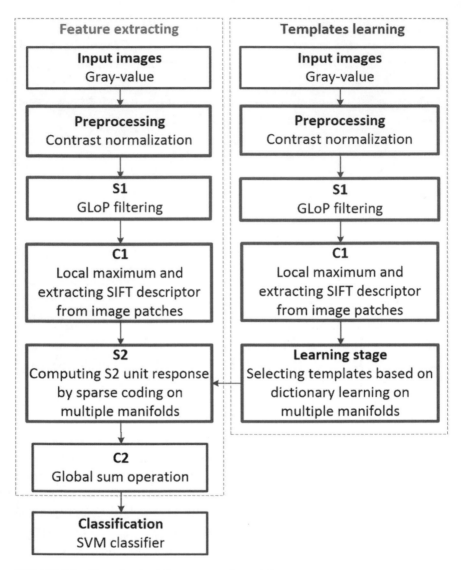

FIGURE 5.2   Flowchart of the proposed model.

5.2.2.2 Gray scale processing

As people have different sensitivity to different colours, we used weighted average method following CCIR 601[29] to transform an input image into a grey one, which can be described as:

$$I = 0.3R + 0.59G + 0.11B \qquad (5.3)$$

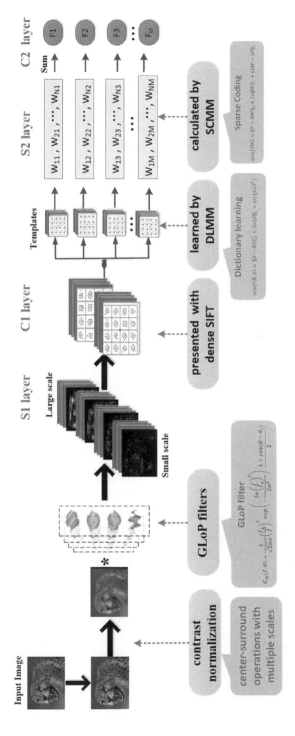

FIGURE 5.3   Framework of the E-HMAX model.

where R, G, and B are respectively the red, green and blue channels of the input image.

5.2.2.3 Contrast normalization

Contrast normalization corresponds to LGN processing using centre-surround operations with multiple scales[30]. For the grey image I, an on-centre, off-surround network performing contrast normalization follows:

$$r_{ij}^{s} = \frac{I_{ij} - \sum_{mn} I_{mn} G_{ijmn}^{s}}{1 + I_{ij} + \sum_{mn} I_{mn} G_{ijmn}^{s}} \tag{5.4}$$

where r_{ij}^{s} represents the response of LGN neuron at position (i,j) and scale s. G_{ijmn}^{s} is a Gaussian spatial kernel for position (i,j) and scale s, which is defined as:

$$G_{ijmn}^{s} = \frac{1}{2\pi\sigma_{s}^{2}} \exp\left[-\frac{(i-m)^{2} + (j-n)^{2}}{2\sigma_{s}^{2}}\right] \tag{5.5}$$

where $\sigma_{s} = [1, 4]$ is the scale variances, and m, n is the size of Gaussian kernel.

5.2.2.4 S1 layer

Visual information in V1 area is analyzed by orientation and frequency bands[31]. Schwartz combined the Fast Fourier Transform (FFT) with log-polar transformation for the visual processing in V1 area and provided a size-invariant mechanism for human vision[32]. Given that, we use GLoP filters[33] to model V1 simple cells instead of Gabor filters. First, the outputs from previous stage are converted to frequency domain through FFT, and then they are convolved with a set of GLoP filters defined by:

$$G_{ik}(f, \theta) = \frac{1}{\sqrt{2\pi}\sigma} \left(\frac{f_{k}}{f}\right)^{2} \exp\left(-\frac{\ln\left(\frac{f}{f_{k}}\right)^{2}}{2\sigma^{2}}\right) \frac{1 + \cos(\theta - \theta_{i})}{2} \tag{5.6}$$

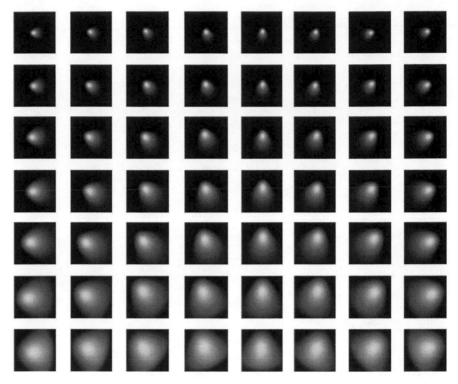

FIGURE 5.4 GLoP filters at eight orientations and frequency bands (wavelet scales).

These filters centre on normalized frequency f_k in the orientation θ_i with the scale parameter σ^{34}. Fig. 5.4 shows the magnitude responses of GLoP filters at eight orientations and eight frequency bands (wavelet scales).

Fig. 5.5 shows the convolution results of the input image (a leopard image) with the GLoP filters presented in Fig. 5.4. Finally, an Inverse Fast Fourier Transform (IFFT) is performed to convert the convolved image back to spatial domain.

Compared with Gabor filters, GLoP filters have an advantage of being symmetric in frequency log scale[34] and are more consistent with human visual system. Better performance can be achieved with more orientations and frequency bands, but it will accordingly increase the time complexity. As a compromise, we only use 4 orientations (0°, 45°, 90°, and 135°) and 8 frequency bands (see Table 4.1), leading to 32 different convolved images for each afferent image.

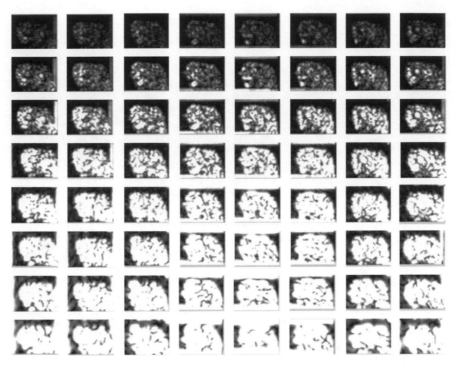

FIGURE 5.5   Convolution results of the leopard image with 64 GLoP filters.

5.2.2.5 C1 layer

C1 units first subsample afferent images from the previous layer using a local maximum pooling from the same scale band (each scale band includes two adjacent filter scales) with the same orientation. The pooling parameters are listed in Table 5.1.

Then SIFT descriptors are extracted from the subsampled images. Each image is densely divided into 16×16 sub-patches, and then SIFT descriptor is extracted on each sub-patch, which will result in a set of image patches which are described with a 128-dimensional vector.

As SIFT descriptor has good tolerance to changes in image translation, scale, and rotation, so SIFT-based image patches can improve the invariance to image rotation and translation change.

5.2.2.6 Template learning stage

The learning stage aims to select a certain number of image templates for S2 layer, which were selected randomly in the HMAX model.

TABLE 5.1 Summary of C1 and GLoP filter parameters

| Scale band S | C1 layer | | GLoP filter |
	Spatial pooling grid (Ns × Ns)	Overlap Δs	f_k
Band 1	8 × 8	4	0.33
			0.22
Band 2	10 × 10	5	0.15
			0.09
Band 3	12 × 12	6	0.07
			0.05
Band 4	14 × 14	7	0.03
			0.02

As most real-world data often lie on multi-manifolds and manifold regularization exploits the geometry of the probability distribution[35], DLMM[36] is used to learn templates from C1 image patches and select representative[37] and meaningful patches. First C1 image patches are extracted from a target set of images. Then topological structure of the multi-manifolds on which image patches reside is constructed by combining sparse representation with k-nearest neighbour algorithm. Assume $X \in \mathbb{R}^{D \times N}$ (D is the feature dimension of X and N is the number of image patches) is the set of all available C1 image patches, the topological structure of the multi-manifolds can be learned by:

$$\min f(U_{i.}^k) = \|X_{.i} - N_k(X_{.i})U_{i.}^k\|^2 + 2\lambda\|U_{i.}^k\|_1 \tag{5.7}$$

where $X_{.i}$ is the ith vector of X, $\mathcal{N}_k(X_{ki}) \in \mathbb{R}^{D \times k}$ represents the k neighbours of $X_{.i}$, and $U_{i.}^k \in \mathbb{R}^{1 \times k}$ is the sparse coefficient of $X_{.i}$ under the base $\mathcal{N}_k(X_{.i})$. Then the coefficient between $X_{.i}$ and $X_{.j}$ can be obtained by:

$$U_{ij} = \begin{cases} U_{ij}^k & X_{.j} \in N_k(X_{.i}) \\ 0 & else \end{cases} \tag{5.8}$$

In Equation (5.7), the coefficient matrix U^k can be obtained by feature-sign search algorithm 39, where λ is set to 0.05 and k is set to 10.

After getting the topological structure U, the set of image templates $T \in \mathbb{R}^{D \times M}$ (M is the number of templates) can be learned by solving the following minimization problem[36]:

$$\min_{T,S} f\,(T,\,S) = \|X - TS\|_F^2 + 2\alpha\|S\|_1 + \eta\,tr\,\{SGS^T\} \qquad (5.9)$$

where α is the sparsity regularization parameter, and η is used to make a tradeoff between reconstruction error and topology preservation. Term $tr\,\{SGS^T\}$ denotes the trace of SGS^T and represents the reservation of local topology. $G \in \mathbb{R}^{N \times N}$ is a Laplacian matrix and can be got by $G = (I - U)(I - U)^T$. Here, parameter α is set to 0.15, and η is set to 0.2.

5.2.2.7 S2 layer

In consideration of the sparse firing properties of V4 cells and manifold way of visual perception, SCMM is used to compute S2 unit responses. For the afferent image patches from C1 layer, we search for a sparse representation[38] by using image templates extracted during learning stage as the dictionary. The corresponding sparse coefficients are exactly the unit response between image patches and templates.

Assume $Y \in \mathbb{R}^{D \times K}$ (K is the number of image patches) is the set of afferent image patches from C1 layer and $T \in \mathbb{R}^{D \times M}$ is the set of image templates, Y can be linearly represented by T:

$$Y \approx TW = \begin{bmatrix} t_{11} & t_{12} & \cdots & t_{1M} \\ t_{21} & t_{22} & \cdots & t_{2M} \\ \vdots & \vdots & \vdots & \vdots \\ t_{D1} & t_{D2} & \cdots & t_{DM} \end{bmatrix} \cdot \begin{bmatrix} w_{11} & w_{12} & \cdots & w_{1K} \\ w_{21} & w_{22} & \cdots & c_{2K} \\ \vdots & \vdots & \vdots & \vdots \\ w_{M1} & w_{M2} & \cdots & c_{MK} \end{bmatrix} \qquad (5.10)$$

where $W \in \mathbb{R}^{M \times K}$ is the sparse coefficient matrix and w_{ij} denotes the unit response between the ith template and jth image patch. Unit response W can be obtained by solving the following problem[36]:

$$\min f\,(W) = \|Y - TW\|_F^2 + 2\alpha\|W\|_1 + \eta\|W - SP\|_2 \qquad (5.11)$$

where $P \in \mathbb{R}^{N \times K}$ is the coefficient matrix by linearly fitting Y with X (image patches for learning templates T), which can be obtained by feature-sign search algorithm[39]. $S \in \mathbb{R}^{M \times N}$ is the sparse coefficient of X under the bases T, which is obtained during the template learning stage.

5.2.2.8 C2 layer

The C2 units are computed by a global summation over all scales and positions of S2 units. For each template, we sum up all S2 unit responses associated with the template. In matrix W, each row corresponds to the responses between a given template and all image patches. So the C2 unit response can be achieved by pooling matrix W with row-wise summation and forms an M-dimensional vector C:

$$C = \left[\sum_{i=1,\dots,K} w_{1i}, \sum_{i=1,\dots,K} w_{2i}, \dots, \sum_{i=1,\dots,K} w_{Mi} \right] \tag{5.12}$$

where M is the number of image templates.

5.3 EXPERIMENTAL RESULTS

In this section, we performed experiments on five benchmark datasets to demonstrate the performance of the proposed model. In Section 5.3.1, experiments were performed on Caltech256 dataset to analyze the effectiveness of each improvement including GLoP filters, SIFT features, and sparse coding for computation of S2 unit response. Section 5.3.2 evaluated the model's robustness to local rotation on Caltech101 dataset. Section 5.3.3 evaluated the model's performance under complex conditions on GRAZ-01 datasets. In Section 5.3.4, experiment was conducted on COIL-20 dataset to show the model's robustness to changes in viewpoint and pose. Last we evaluated our model on Scene13 dataset. The extracted C2 features were fed to a linear support vector machine (libSVM[40]) to perform the classification task. We run all experiments 10 times and averaged the results. Classification rate and receiver operating characteristic (ROC) curve[41] were used to measure the classification performance.

5.3.1 Effectiveness analysis of GLoP filters, SIFT features, SCMM, and DLMM

In this section, we choose four object categories (bat, bear, canoe, and watermelon) as positive set and backgrounds as negative set. The four categories consist of images which change under various position, size, viewpoint, and lighting conditions. Twenty images were used for

FIGURE 5.6 Exemplars from Caltech256 dataset. The categories are respectively bat, bear, canoe, and watermelon from top to bottom.

training and 30 for testing. Fig. 5.6 shows some exemplars from the four categories.

Fig. 5.7 shows the comparison between Gabor filters and GLoP filters with different numbers of features on the selected object categories. GLoP filters performed better than the standard Gabor filters for most categories (bat, bear, and watermelon), only showing a litter worse on canoe when there are 50 and 100 features. So GLoP filters have an advantage over Gabor filters and using GLoP filters in S1 layer can improve the performance of the model. As SCMM for S2 unit response calculation was closely related to DLMM for template selection, testing them individually was not necessary and meaningful.

Fig. 5.8 presents the comparison between Euclidean distance and SCMM in computing S2 unit response. Results show that SCMM significantly outperforms Euclidean distance when computing S2 unit response. As illustrated in Fig. 5.9, SIFT-based image patches also show some advantages over pixel-based patches. Thus, it can be seen that our modifications to the HMAX model are not only biologically feasible but also achieve better performance.

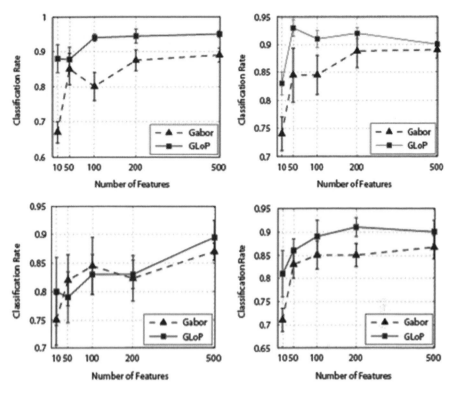

FIGURE 5.7 Comparison between Gabor and GLoP filters on bat, bear, canoe, and watermelon.

5.3.2 Evaluation of local rotation

To evaluate the influence of local rotation on the recognition performance, experiment was performed on Caltech101 dataset[42] which is available at http://www.vision.caltech.edu/Image_Datasets/Caltech101/. As described in[43], we selected four object categories (airplanes, cups, laptops, and guitars) as positive sets and backgrounds as negative set. Fifteen images were randomly chosen as training examples and their rotational versions with increasing angle as testing samples (as shown in Fig. 5.10). Only one orientation of GLoP filters and 200 features were used in the E-HMAX model. We compare our method with the HMAX model and two extended models (SSMF[44] and OGHM-HMAX[43]) which are derived from the HMAX model and aim to improve the rotational invariance of the HMAX model.

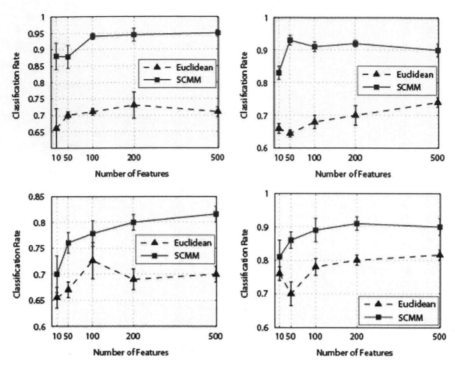

FIGURE 5.8 Comparison between Euclidean distance and SCMM for the computation of S2 unit response on bat, bear, canoe, and watermelon.

As shown in Fig. 5.11, E-HMAX shows a significant improvement over the standard HMAX, especially with large rotational angles. When the rotational angle increases from 30° to 45°, there is a significant decline in the classification rate of HMAX, SSMF, and OGHM-HMAX. Relatively speaking, our model doesn't show an obvious decrease with the increase of rotational angle. This suggests that the E-HMAX is not so sensitive to rotational transformation as the other models. In general, E-HMAX model achieves the best performance under condition of rotation and is more robust against rotational changes than the contrast models.

5.3.3 GRAZ-01 dataset

The GRAZ-01 dataset[46] (Fig. 5.12) available at http://www.emt.tugraz.at/~pinz/data/GRAZ_01/ is a challenging dataset with high intra-class variability on highly cluttered backgrounds. As in[46], 100 samples were randomly chosen for training and 50 other samples were used for

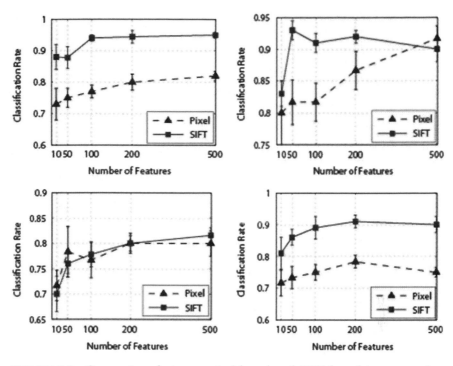

FIGURE 5.9 Comparison between pixel-based and SIFT-based image patches in C1 units on bat, bear, canoe, and watermelon.

testing. The proposed model was compared with some feature extraction methods: SIFT, SM, HMAX, Ghodrati's method in Ghodrati et al.[47], and EBIM in Huang et al[24]. SIFT and SM were local feature extraction methods. HMAX was used to provide the baseline performance for the experiment; Ghodrati's method and EBIM were prevalent enhanced models original from HMAX. The comparison results are presented in Table 5.2 and Fig. 5.13. It can be seen from Table 5.2 that E-HMAX exhibits the best performance in all cases. ROC curve (Fig. 5.13) shows that E-HMAX significantly outperforms HMAX and SIFT for both bikes and persons. The ROC curves of E-HMAX are comparable to those of EBIM. When the FPR (false positive rate) is greater than 0.18, E-HMAX has a slightly lower true positive rate; however, when FPR is less than 0.18, E-HMAX greatly outperforms EBIM model. This experiment implies the advantage of our method in a challenge recognition task (Fig. 5.13, Table 5.2).

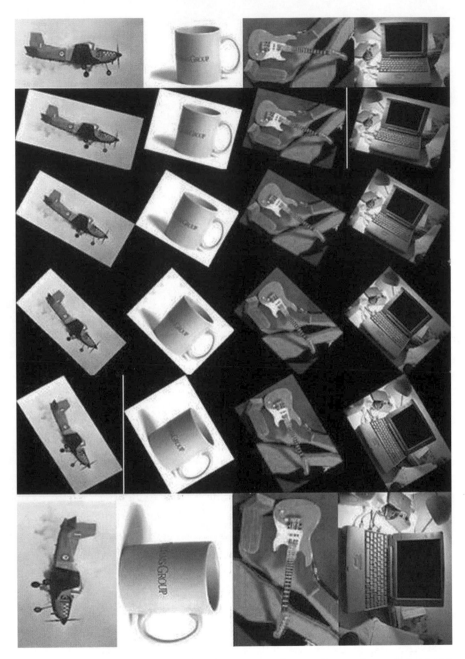

FIGURE 5.10 Sample images for the rotational experiment from Caltech101 dataset. First column: the training images. Second to sixth columns: images with various rotation angles (15°, 30°, 45°, 60°, and 90°).

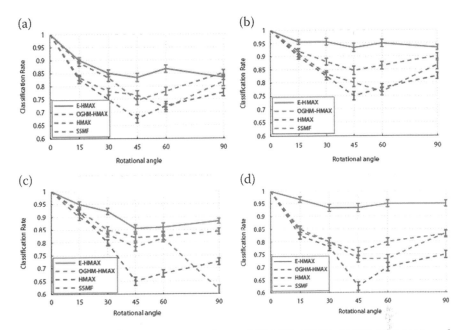

FIGURE 5.11   Performance comparison of HMAX, SSMF, OGHM-HMAX, and E-HMAX in local rotation on four object categories. (a) Airplanes; (b) cups; (c) guitars; (d) laptops.

FIGURE 5.12   Sample images from GRAZ-01 dataset. From top to bottom the categories are respectively bikes, person, and background.

5.3.4 COIL-20 dataset

COIL-20 dataset available at http://www.cs.columbia.edu/ CAVE/ software/softlib/coil-20.php contains 1440 images of 20 different objects (Fig. 5.14(a)). Each object includes 72 images taken at pose interval

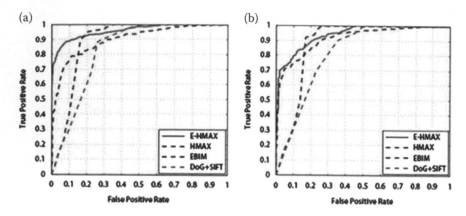

FIGURE 5.13   ROC curves of several methods including SIFT and DoG interest point detection (DoG+SIFT) reported in Opelt et al.[46], HMAX and EBIM reported in Huang et al.[24]. (a) Bikes; (b) person.

TABLE 5.2   Experimental results of several methods on GRAZ-01 dataset. Results of first two rows are reported in Opelt et al.[46] and results of the third and fifth row are reported in Ghodrati et al.[47] and Huang et al.[24]

Method	Bikes		Person	
	[a]EER	[b]AUC	EER	AUC
SIFT[46]	78.0	86.5	76.5	80.8
SM[46]	83.5	89.6	56.5	59.1
HMAX	75.5	85.2	74.8	81.5
Ghodrati et al.[47]	80.2	88.5	84	90.8
EBIM[24]	84.1	90.5	86.0	91.8
E-HMAX	86.3	93.4	86.1	92.5

Notes
[a] EER: Detection rate at equal-error-rate of the ROC curve.
[b] AUC: Area under the ROC curve.

FIGURE 5.14   Exemplars from COIL-20 dataset. Left is sample images of 20 objects and right is example sequence of the first object.

of 5 degrees[48], and some samples of the first object are shown in Fig. 5.14(b). In this experiment, 24 and 36 images were respectively used for training and the remaining images for testing. This experiment was performed on a desktop with an Intel Core i7–4770 CPU 3.4 GHz and 16 G RAM. As deep learning networks have achieved outstanding performance on object recognition, we compared the proposed model with a Deep Neural Network (sparse Auto-encoder: SAE)[49]. Meanwhile, HMAX was also used to provide the baseline performance. The experimental results including classification accuracy and computational speed are presented in Table 5.3. For SAE network, each image was resized to 64×64 and organized into a vector of 4096. The SAE network consists of an input layer with 4096 input units, two sparse auto-encoder layers with 1000 hidden units, and a softmax classifier. Model parameters are set as $\lambda = 0.003$, $\beta = 3$, and $\rho = 0.1$.

The proposed model greatly outperformed the HMAX model both in classification accuracy and processing speed, especially when there are small number of training samples. It indicates that our model is more robust to changes in viewpoint and pose than the standard HMAX model. When there are 24 training images, our model outperforms SAE network; but when there are 36 training images, our model performs a little worse than SAE. Generally speaking, E-HMAX and SAE achieve equivalent classification accuracy in this task. Though deep learning networks have made great achievements in visual recognition, they may not be of much advantage with a small amount of training samples. However, in view of computational efficiency, SAE is more time-consuming and spends approximately 90 times as much time as spent on the E-HMAX model. In general, our method can achieve satisfactory performance both on classification accuracy and processing speed in this experiment especially when there are small numbers of training samples.

TABLE 5.3 Classification performances of E-HMAX, HMAX, and SAE

Methods	Accuracy (%)		Processing Time(s)
	24	36	
E-HMAX	85.2	90.3	246.6
HMAX	79.3	88.3	651.9
SAE	83.4	90.8	21863.8

	1	2	3	4	5	6	7	8	9	10	11	12	13	14	15	16	17	18	19	20
1	85.42	0.00	0.00	0.00	0.00	0.00	0.00	0.00	0.00	0.00	0.00	0.00	0.00	0.00	0.00	0.00	0.00	0.00	0.00	0.00
2	0.00	41.67	0.00	0.00	0.00	4.17	0.00	0.00	0.00	0.00	0.00	0.00	0.00	0.00	0.00	2.08	0.00	0.00	0.00	0.00
3	0.00	0.00	47.92	0.00	27.08	37.50	0.00	0.00	0.00	0.00	0.00	0.00	0.00	0.00	0.00	0.00	0.00	0.00	0.00	0.00
4	0.00	0.00	0.00	95.83	0.00	0.00	0.00	0.00	0.00	0.00	0.00	0.00	0.00	0.00	0.00	0.00	0.00	0.00	0.00	0.00
5	0.00	16.67	2.08	0.00	43.75	8.33	18.75	0.00	0.00	0.00	0.00	0.00	0.00	0.00	0.00	0.00	0.00	0.00	0.00	0.00
6	0.00	0.00	4.17	0.00	0.00	33.33	0.00	0.00	0.00	0.00	0.00	0.00	0.00	0.00	0.00	0.00	0.00	0.00	0.00	0.00
7	0.00	6.25	4.17	0.00	4.17	0.00	81.25	0.00	6.25	0.00	0.00	0.00	0.00	0.00	0.00	0.00	0.00	0.00	0.00	0.00
8	0.00	0.00	0.00	0.00	0.00	0.00	0.00	100.00	0.00	0.00	0.00	0.00	0.00	0.00	0.00	0.00	0.00	0.00	0.00	0.00
9	0.00	0.00	5.25	0.00	18.75	4.17	0.00	0.00	93.75	0.00	0.00	0.00	0.00	0.00	0.00	0.00	0.00	0.00	0.00	0.00
10	0.00	0.00	0.00	0.00	0.00	0.00	0.00	0.00	0.00	100.00	0.00	0.00	0.00	0.00	0.00	0.00	0.00	0.00	0.00	0.00
11	0.00	25.00	0.00	0.00	0.00	0.00	0.00	0.00	0.00	0.00	91.67	0.00	8.33	0.00	0.00	0.00	0.00	0.00	0.00	0.00
12	0.00	0.00	0.00	0.00	0.00	0.00	0.00	0.00	0.00	0.00	0.00	100.00	0.00	0.00	0.00	0.00	0.00	0.00	0.00	0.00
13	0.00	0.00	0.00	4.17	0.00	0.00	0.00	0.00	0.00	0.00	0.00	0.00	91.67	0.00	0.00	0.00	0.00	0.00	0.00	0.00
14	14.58	10.42	0.00	0.00	0.00	0.00	0.00	0.00	0.00	0.00	8.33	0.00	0.00	100.00	0.00	0.00	0.00	0.00	0.00	0.00
15	0.00	0.00	0.00	0.00	0.00	0.00	0.00	0.00	0.00	0.00	0.00	0.00	0.00	0.00	100.00	0.00	0.00	0.00	0.00	0.00
16	0.00	0.00	0.00	0.00	0.00	0.00	0.00	0.00	0.00	0.00	0.00	0.00	0.00	0.00	0.00	97.92	0.00	0.00	0.00	0.00
17	0.00	0.00	0.00	0.00	0.00	0.00	0.00	0.00	0.00	0.00	0.00	0.00	0.00	0.00	0.00	0.00	100.00	0.00	0.00	0.00
18	0.00	0.00	0.00	0.00	0.00	0.00	0.00	0.00	0.00	0.00	0.00	0.00	0.00	0.00	0.00	0.00	0.00	100.00	0.00	0.00
19	0.00	0.00	35.42	0.00	6.25	12.50	0.00	0.00	0.00	0.00	0.00	0.00	0.00	0.00	0.00	0.00	0.00	0.00	100.00	0.00
20	0.00	0.00	0.00	0.00	0.00	0.00	0.00	0.00	0.00	0.00	0.00	0.00	0.00	0.00	0.00	0.00	0.00	0.00	0.00	100.00

FIGURE 5.15 Confusion matrix of the E-HMAX model on COIL-20 dataset.

Fig. 5.15 presents the confusion matrices of the proposed model for 24 training samples. Relatively speaking, objects 2, 3, 5, and 6 have lower classification accuracy. Some misclassified images from poor-performing objects are shown in Fig. 5.16. In general, there are two reasons leading to misclassifications: one reason is that the involved objects have complex shapes such as objects 2 and 7, and there is great variation among different viewpoints; the other is that some objects are very similar in appearances, such as objects between 5 and 9, objects among 3, 6, and 19.

5.3.5 Scene13 dataset

To evaluate our modification in S1 layer, we compared our model with three models which involve RFs model in S1 layer: the standard HMAX (Gabor filters), HMAX computed by DOG (denoted as HMAX+DOG), and HMAX computed by OGHM (denoted as OGHM-HMAX). We performed the experiment on Scene13 dataset. As in Lu et al.[43], we randomly selected 15 positive training samples from each scene category and 15 negative training samples from backgrounds. 50 positive samples and 50 negative samples were chosen for testing. Fig. 5.17 shows some sample images from Scene13 dataset.

The experimental results are presented in Fig. 5.18 and the mean accuracy is shown in Table 5.4. E-HMAX has significantly outperformed HMAX and HMAX+DOG in all cases. Our model also outperforms the OGHM-HMAX for most categories except kitchen

FIGURE 5.16 Some misclassified images from poor performing objects on COIL-20 dataset.

FIGURE 5.17 Sample images from Scene13 dataset: they are respectively bedroom, suburb, kitchen, living room, office, coast, forest, highway, inside of city, mountain, open country, street, and tall building. The last one is a background image.

and office. In general, the E-HMAX model demonstrated the best performance in this experiment and it is a powerful RFs model based on HMAX in scene recognition.

5.4 DISCUSSION

5.4.1 Computer vision perspective on the E-HMAX

An image of the visual world is created in the retina and then transferred to the LGN, which is a relay centre for the visual pathway. LGN forwards the visual information to the visual cortex which consists of V1, V2, V4, and inferior temporal (IT) areas[50] for higher-level processing.

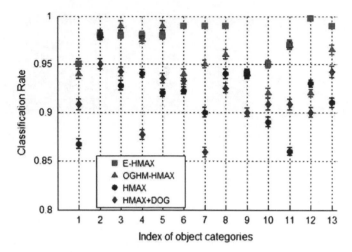

FIGURE 5.18   Results of HMAX-based models on Scene13 dataset. The index of object categories (1–13) respectively represent bedroom, suburb, kitchen, living room, office, coast, forest, highway, inside of city, mountain, open country, street, and tall building.

TABLE 5.4   Comparison of classification results on Scene 13 dataset

Methods	Mean accuracy (%)
HMAX-DOG	91.38 ± 2.53
HMAX	91.62 ± 3.20
OGHM-HMAX	95.69 ± 2.31
E-HMAX	97.68 ± 1.79

The proposed model exactly simulates the visual processing of human visual system and each improvement originates from research findings in neurophysiological and cognitive neuroscience of human visual system.

We used contrast normalization to simulate retina and LGN processing. In the LGN, most neurons have approximately circular centre-surround organization in the spatial domain[51]. LGN contains ON cells and off cells. ON cells obey membrane equations and interact via on-centre, off-surround interactions which normalize local image contrasts. Cells in the OFF channel interact via an off-centre, on-surround interaction[52]. So contrast normalization can well describe the neural response of LGN neurons.

At S1 layer, input images were convolved with GLoP filters rather than Gabor filters. Related research shows GLoP filters are more consistent with measurement of the primate visual systems, which have neural response symmetric on the log frequency scale[53,54]. Meanwhile, experimental results on Caltech256 dataset have shown the advantage of GLoP filters over Gabor filters.

At C1 layer, image patches were presented with SIFT descriptor rather than pixel-based value. SIFT features possess similar properties with neurons in IT cortex which account for object recognition in human vision[55].

5.4.2 Invariance of local rotation

One prominent advantage of the E-HMAX model is the improvement of the invariance to the rotational transformations. Experimental results show our model not only shows great improvement over the HMAX model but also outperforms SURF-HMAX and OGHM-HMAX (Fig. 5.11). E-HMAX features share good invariance to translation, scaling, and rotation due to the good tolerance to image translation, scale, and rotation of SIFT descriptor. We use dense SIFT algorithm rather than original SIFT for two reasons. One is that dense SIFT is a fast dense version of the SIFT. The other is that dense SIFT extracts SIFT features on all positions of the interested patches and is more suitable for object recognition than SIFT while SIFT and SURF extract features are based on interest points.

5.4.3 Limitations and possible improvement

There seem to be two directions that could be improved further. First, the major limitation of our model remains its processing speed. Though we have made some efforts to improve the processing speed, such as selecting fewer Gabor filter bands, the processing speed is still too slow for a real-time application and there is still work to do on the processing speed.

Second, though our model shows some robustness to viewpoint in COIL-20 experiment, it has not specialized learning mechanism providing view-invariant presentation. Without such mechanism, E-HMAX does not possess the key properties of many neurons in the IT visual cortex[45].

5.5 CONCLUSIONS

In this chapter, we present a simple yet powerful biologically inspired model to improve the performance of the HMAX model. Contrast normalization is made to simulate the processing of human retina and LGN and generate more robust visual information before feature extraction. GLoP filters are used to simulate the properties of V1 simple cells and increase the robustness to rotation and zoom. We integrate SIFT descriptors into the HMAX model to improve its invariance to rotation. Considering the sparse firing properties of V4 neurons and manifold way of visual perception, sparse coding on multi-manifolds is used to compute the simple cell response of V4 areas instead of Euclidean distance.

We have conducted experiments on five benchmark databases to demonstrate the effectiveness of the proposed model. Results show that our model greatly outperformed the HMAX model in a variety of object classification tasks. In contrast to the HMAX model, its tolerance to rotational transformation has been improved to some degree, especially local transformation. Though E-HMAX does not perform the best in all cases, it is comparable to some state-of-the-art methods.

REFERENCES

1. Li, H., Li, H., Wei, Y., Tang, Y., Wang, Q. 2014. Sparse-based neural response for image classification. *Neurocomputing* 144:198–2077.
2. Yu, J., Tao, D., Rui, Y., Cheng, J. 2013. Pairwise constraints based multiview features fusion for scene classification. *Pattern Recognition* 46(2): 483–496.
3. Sang, J., Xu, C., Liu, J. 2012. User-aware image tag refinement via ternary semantic analysis. *IEEE Transactions on Multimedia* 14(3):883–895.
4. Sang, J., Fang, Q., Xu, C. 2017. Exploiting social-mobile information for location visualization. *ACM Transactions on Intelligent Systems and Technology (TIST)* 8(3):39.
5. Tan, M., Hu, Z., Wang, B., Zhao, J., Wang, Y. 2016. Robust object recognition via weakly supervised metric and template learning. *Neurocomputing* 181:96–107.
6. Tan, M., Wang, B., Wu, Z., Wang, J., Pan, G. 2016. Weakly supervised metric learning for traffic sign recognition in a lidar-equipped vehicle. *IEEE Transactions on Intelligent Transportation Systems* 17(5): 1415–1427.

7. Fukushima, K. 1980. Neocognitron: A self-organizing neural network model for a mechanism of pattern recognition unaffected by shift in position. *Biological Cybernetics* 36(4):193–202.
8. Serre, T., Wolf, L., Bileschi, S., Riesenhuber M. 2007. Robust object recognition with cortex-like mechanisms. *IEEE Transactions on Pattern Analysis and Machine Intelligence* 29(3):411–426.
9. Deng, L., Wang, Y., Liu, B., et al. 2018. Biological modelling of human visual system for object recognition using GLoP filters and sparse coding on multi-manifolds. *Machine Vision and Applications* 29(6):965–977.
10. Lee, H., Grosse, R., Ng, A. 2009. Convolutional deep belief networks for scalable unsupervised learning of hierarchical representations. Proceedings of the 26th Annual International Conference on Machine Learning, ICML 2009, Montreal, Quebec, Canada, ACM.
11. Krizhevsky, A., Sutskever, I., Hinton, G. E. 2012. Imagenet classification with deep convolutional neural networks. *Advances in Neural Information Processing Systems* 25:1097–1105.
12. Kheradpisheh, S., Ghodrati, M., Ganjtabesh, M., Masquelier, T. 2015. Deep Networks Resemble Human Feed-forward Vision in Invariant Object Recognition. *arXiv preprint arXiv:1508.03929.*
13. Girshick, R., Donahue, J., Darrell, T., Malik, J. 2014. Rich feature hierarchies for accurate object detection and semantic segmentation. In Proceedings of the IEEE conference on computer vision and pattern Recognition: 580–587.
14. Yu, J., Zhang, B., Kuang, Z., Lin, D., Fan, J. 2017. iPrivacy: image privacy protection by identifying sensitive objects via deep multi-task learning. *IEEE Transactions on Information Forensics and Security* 12(5):1005–1016.
15. Yu, J., Yang, X., Gao, F., Tao, D. 2016. Deep multimodal distance metric learning using click constraints for image ranking. *IEEE Transactions on Cybernetics PP* 99:1–11.
16. Wu, W., Qiao, H., Chen, J., Yin, P., Li, Y. 2016. Biologically inspired model simulating visual pathways and cerebellum function in human-Achieving visuomotor coordination and high precision movement with learning ability. *arXiv preprint arXiv:1603.02351.*
17. Cadieu, C., Kouh, M., Pasupathy, A., Connor, C. E., Riesenhuber M., Poggio T. 2007. A model of V4 shape selectivity and invariance. *Journal of Neurophysiology* 98:1733–1750.
18. Weng, D., Wang, Y., Gong, M., Tao, D., Wei, H. 2015. DERF: Distinctive efficient robust features from the biological modelling of the P ganglion cells. *IEEE Transactions on Image Processing* 24(8):2287–2302.
19. Grossberg, S., Hong, S. 2006. A neural model of surface perception: Lightness, anchoring, and filling-in. *Spatial Vision* 19:263–321.
20. Olshausen, B. A., Field, D. J. 1996. Emergence of simple-cell receptive field properties by learning a sparse code for natural images. *Nature* 381:607–609.

21. Carlson, E. T., Rasquinha, R. J., Zhang, K., Connor, C. E. 2011. A sparse object coding scheme in area V4. *Current Biology* 21:288–329.

22. Quiroga, Q. R., Reddy, L., Kreiman, G., Koch, C., Fried, I. 2005. Invariant visual representation by single neurons in the human brain. *Nature* 435:1102–1107.

23. Hu, X., Zhang, J., Li, J., Zhang, B. 2014. Sparsity-regularized HMAX for visual recognition. *PloS One* 9(1):e81813.

24. Huang, Y., Huang, K., Tao, D., Tan, T., Li, X. 2011. Enhanced biologically inspired model for object recognition. *IEEE Transactions on Systems, Man, and Cybernetics, Part B (Cybernetics)* 41(6):1668–1680.

25. Liu, W., Zha, Z. J., Wang, Y., Lu, K., Tao, D. 2016. P-laplacian regularized sparse coding for human activity recognition. *IEEE Transactions on Industrial Electronics* 63(8):5120–5129.

26. Yu, J., Rui, Y., Tao, D. 2014. Click prediction for web image reranking using multimodal sparse coding. *IEEE Transactions on Image Processing* 23(5):2019–2032.

27. Seung, H. S., Lee, D. D. 2000. The manifold ways of perception. *Science* 290(5500):2268–2269.

28. J. Weng, N. Ahuja, T. S. Huang. 1993. Learning recognition and segmentation of 3-D objects from 2-D. In Proc. IEEE 4th Int'l Conf. Computer Vision: 121–128.

29. Sector, I. T. U. R. 1995. Studio encoding parameters of digital television for standard 4:3 and wide-screen 16:9 aspect ratios. International Telecommunication Union Radiocommunications Sector (ITU-R) BT 601–605.

30. Grossberg, S., Huang, T. R. 2009. ARTSCENE: a neural system for natural scene classification. *Journal of Vision* 9 (4): 1–19.

31. De Valois, R. L., Yund, E. W., Hepler, N. 1982. The orientation and direction selectivity of cells in macaque visual cortex. *Vision Research* 22: 531–544.

32. Schwartz, E. L. 1981. Cortical anatomy and size invariance, and spatial frequency analysis. *Vision Research* 18:24–58.

33. Guyader, N., Chauvin, A., Massot, C., Hérault, J., Marendaz, C. 2006. A biological model of low-level vision suitable for image analysis and cognitive visual perception. *Perception* 35(1):56.

34. Benoit, A., Caplier, A., Durette, B., Herault, J. 2010. Using human visual system modelling for bio-inspired low level image processing. *Computer Vision and Image Understanding* 114 (7):758–773.

35. Liu, T., Tao, D. 2016. Classification with noisy labels by importance reweighting. *IEEE Transactions on Pattern Analysis and Machine Intelligence* 38(3):447–461.

36. Liu, B., Wang, Y., Zhang, Y., Shen, B. 2013. Learning dictionary on manifolds for image classification. *Pattern Recognition* 46(7): 1879–1890.

37. Tao, D., Li, X., Wu, X., Maybank, S. J. 2009. Geometric mean for subspace selection. *IEEE Transactions on Pattern Analysis and Machine Intelligence* 31(2):260–274.
38. Yu J., Rui Y., Tao D. 2014. Click prediction for web image reranking using multimodal sparse coding. *IEEE Transactions on Image Processing* 23(5):2019–2032.
39. Lee, H., Battle, A., Raina, R., Ng A. Y. 2006. Efficient sparse coding algorithms. *Advances in Neural Information Processing Systems* 19:801–808.
40. Chang, C. C., Lin, C. J. 2011. LIBSVM: a library for support vector machines. *ACM Transactions on Intelligent Systems and Technology (TIST)* 2(3):1–27.
41. Park, S. H., Goo, J. M., Jo, C. H. 2004. Receiver operating characteristic (ROC) curve: Practical review for radiologists. *Korean Journal of Radiology* 5(1):11–18.
42. Li, F. F., Fergus, R., Perona, P. 2007. Learning generative visual models from few training examples: An incremental bayesian approach tested on 101 object categories. *Computer Vision and Image Understanding* 106(1):59–70.
43. Lu, Y. F., Zhang, H. Z., Kang, T. K., Choi, I. H., Lim, M. T. 2014. Extended biologically inspired model for object recognition based on oriented Gaussian–Hermite moment. *Neurocomputing* 139:189–201.
44. Jiang, L. Y. 2014. *Study on Bio-inspired invariant feature representation of image*, MS Thesis, Qingdao: China University of Petroleum.
45. Robinson, L., Rolls, E. T. 2015. Invariant visual object recognition: biologically plausible approaches. *Biological Cybernetics* 109 (4–5):505–535.
46. Opelt, A., Pinz, A., Fussenegger, M., Auer, P. 2006. Generic object recognition with boosting. *IEEE Transactions on Pattern Analysis and Machine Intelligence* 28 (3):416–431.
47. Ghodrati, M., Khaligh-Razavi, S. M., Ebrahimpour, R., Rajaer, K., Pooyan, M. 2012. How can selection of biologically inspired features improve the performance of a robust object recognition model? *PloS One* 7 (2):e32357.
48. Zhai, D., Li, B., Chang, H., Shan, S., Chen, X., Gao, W. 2010. Manifold alignment via corresponding projections. In: *BMVC*.
49. Liu, W., Ma, T., Tao, D., You, J. 2016. HSAE: A Hessian regularized sparse auto-encoders. *Neurocomputing* 187:59–65.
50. Yin, P., Qiao, H., Wu, W., Qi, L., Li, Y., Zhong, S., Zhang, B. 2016. A novel biologically mechanism-based visual cognition model–automatic extraction of semantics, formation of integrated concepts and re-selection features for ambiguity. *arXiv preprint arXiv:1603.07886*.
51. Lindeberg, T. 2013. A computational theory of visual receptive fields. *Biological Cybernetics* 107 (6):589–635.
52. Bhatt, R., Carpenter, G. A., Grossberg, S. 2007. Texture segregation by visual cortex: Perceptual grouping, attention, and learning. *Vision Research* 47:3173–3211.

53. Field, D. J. 1987. Relations between the statistics of natural images and the response properties of cortical cells. *JOSA A.* 4 (12):2379–2394.
54. Kovesi, P. 1999. Image features from phase congruency. *Videre: Journal of Computer Vision Research* 1 (3):1–26.
55. Serre, T., Kouh, M., Cadieu, C., Knoblich, U., Kreiman, G., Poggio, T. 2005. *A theory of object recognition: computations and circuits in the feedforward path of the ventral stream in primate visual cortex.* Massachusetts Institute of Technology Cambridge MA Center for Biological and Computational Learning.

## CHAPTER 6

Increment learning and rapid retrieval of visual information based on pattern association memory

6.1 INTRODUCTION

Humans can effectively learn to predict and recognize objects in such a rapid-changing world without rapid forgetting what they have known. This outstanding ability is very essential to our daily life[1]. How visual information is represented in the brain and how to retrieve it when needed have drawn great attention in the past few years. Understanding how the brain performs the cognitive functions, such as object recognition and memory recall, requires close associations between the neural data and the computational models of how these cognitive problems are solved[2]. Discovering the solution to this key issue is of great significance not only to understand ourselves but also to develop object recognition applications[3].

DOI: 10.1201/9781003281641-6

In the past few decades, much progress has been made in the fields of cognitive neuroscience and brain modelling[4]. Many neural models have been proposed to simulate how the cerebral cortex conducts the complex and necessary functions in memory. Achieving artificial intelligence requires these models to be capable of learning and remembering many different tasks[5]. However, most of these memory models[4–7] are conceptual and principles models, which are not very easy to be applied to practical applications. Moreover, many neural networks do encounter serious catastrophic forgetting, especially when they attempt to rapidly react to a constantly changing world, which is unacceptable for both a memory model and practical engineering applications[8]. These models include the self-organizing map[9], competitive learning, neocognitron, back propagation, and Bayesian models[10]. So how to learn new information rapidly without interfering with previously stored information is a key issue to be solved.

In the past few years, many approaches have been proposed in the field of image retrieval, including multi-view hypergraph-based learning (MHL)[11], image ranking using user clicks and visual features[12], multimodal sparse coding[13], and multi-view locality-sensitive sparse retrieval[14], and so on[15–17]. Though these methods show excellent performance in image search and retrieval, they are not from the perspective of brain memory modelling and are not really biological plausible. In recent years, deep learning has proven successful in image search and retrieval. For example, to achieve fast and accurate detection of large numbers of object classes, Yu et al.[18] developed a hierarchical deep multi-task learning algorithm to learn more representative deep CNNs and more discriminative tree classifiers. In[19], Yu et al. also proposed a novel deep multimodal distance metric learning (Deep-MDML) method for image ranking. In addition, Hong et al.[20] proposed a novel pose recovery method using multimodal deep autoencoder. However, these deep models are large neural networks and require a large number of training data, which is not easily satisfied under practical conditions[15].

Associative memory has been an active research topic for over 50 years and is still active in the fields of cognitive neuroscience and neural networks[21]. Pattern associators play important role in the outputs of the visual system to the learning systems in the orbitofrontal cortex. They also work throughout the cerebral cortex, visual memory recall, and the short-term memory systems[22]. Given that, the principle of pattern associators can be applied to the storage and recall of visual memory. Rolls and Treves presented a prototypical pattern association network and accounted for the operation of pattern associators in the Appendices of Rolls and Treves[23]. However, the units in the network are just binary ones and the network is only simulated by simple binary pattern association. Moreover, the network only uses a simple Hebbian learning rule, which has some limitations such as myopia, greed, and local optimum, and is not good at dealing with longer-term, larger-scale problems. While error-driven learning can just overcome these limitations[23]. So combining both forms of learning together will provide better performance.

To alleviate the limitations of the pattern association network and provide an effective method for memory modelling, we present an Increment Pattern Association Memory Model (IPAMM) for the storage and retrieval of visual information based on the research findings of cognitive neuroscience[24]. The Leabra learning mechanism[25] is introduced into our model instead of the pure Hebbian learning rule. By assigning individual synapse weight to different pattern categories, increment learning can be achieved and catastrophic interference can be avoided at the same time. Unlike the BP neural networks and deep learning networks, the structure of IPAMM is very simple and only includes two layers. Meanwhile, the proposed model can receive real-value visual information as input patterns (conditioned stimulus). To demonstrate the universality of the proposed model under various image representation methods, we extract image features based on three prevailing methods: ScSPM[26], Sparse-HMAX[27], and deep convolutional neural network (CNN)[28].

The rest of this chapter is organized as follows. We briefly introduce the fundamental principle of pattern association memory in Section 6.2. Section 6.3 gives a detailed description of the proposed model, including feature-extraction methods, learning mechanism, and the process of recall. Section 6.4 presents the experimental results and analysis on three benchmark datasets. Finally, we give the discussion and conclusions in Sections 6.5 and 6.6.

6.2 PATTERN ASSOCIATION MEMORY

A basic function of many neural networks is to associate a first stimulus (conditioned stimulus) with a second (unconditioned stimulus) which

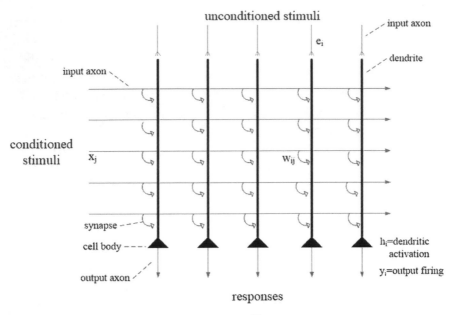

FIGURE 6.1   A pattern association memory[22].

appears almost simultaneously and to retrieve the second when the first is presented. This operation can be achieved by a pattern association network as depicted in Fig. 6.1. Table 6.1 lists some important notations used in this chapter.

TABLE 6.1   List of important notations

Notations	Description
w	Synaptic weight matrix
α	Learning rate
x_j	Presynaptic firing for the jth axon
y_i	postsynaptic activity on the neuron i
h_i	Activation of a neuron i
θ	Inhibitory threshold allowing the top k units to keep active
λ	Weight coefficient adjusting the proportion between CHL learning and Hebbian learning.
D	Dimension of image features (Number of input axons)
N	Number of object classes
M	Number of training samples

As illustrated in Fig. 6.1, the unconditioned stimulus predominates in producing firing (activity) of the output neurons (y_i for the ith neuron). The conditioned stimulus vector x (x_j for the jth axon) is used through synapse w_{ij} to the output neurons. The synapse weight w_{ij} is modified through Hebb learning rule[29], which can be described as:

$$\Delta w_{ij} = \alpha y_i x_j \tag{6.1}$$

where Δw_{ij} is the update of the synaptic weight w_{ij} and α represents the learning rate.

When a conditioned stimulus is presented, the activation h_i of a neuron i can be achieved by accumulating all the activations generated by each active neuron x_j and the corresponding weight w_{ij}, which can be expressed as:

$$h_i = \sum_{j=1}^{D} x_j w_{ji} \tag{6.2}$$

where D is the number of input axons.

Then activation h_i will be converted into firing y_i:

$$y_i = f(h_i) \tag{6.3}$$

where f is the activation function and can take many forms.

6.3 INCREMENT PATTERN ASSOCIATION MEMORY MODEL (IPAMM)

The proposed model includes stages of learning and recall, as depicted in Fig. 6.2. The target of learning stage is to obtain a synapse matrix which can best describe the distribution properties of the images from each class. The Leabra learning framework is used to learn the synapse matrix in our model. When recalling, we first calculate the neural activation of the test image for all classes and then search for the class that best fits the test image.

6.3.1 Feature extraction

To demonstrate the performance of our model under various visual representation algorithms, we apply three prevalent methods for representing

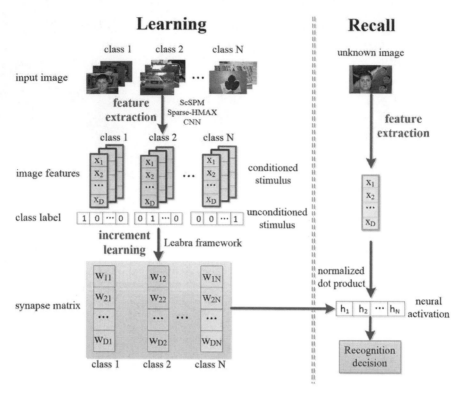

FIGURE 6.2   Flow chart of the proposed model.

the object images: ScSPM, Sparse-HMAX, and deep convolutional neural network (CNN). ScSPM[26] is a spatial pyramid matching method based on sparse coding[30] with high-dimensional features of 21,504 dimensions. Sparse-HMAX[27] is a biologically inspired model which mimics object recognition in visual cortex with lower-dimensional features (200 dimensions in this chapter). Convolutional neural networks have achieved many practical successes and have been widely used in many fields of computer vision. As described in Simonyan and Zisserman[28], the input to the CNN is RGB image with a fixed size of 224×224. There are 19 weight layers (16 convolutional layers and 3 FC layers) in the network and all hidden layers are equipped with rectification (ReLU) non-linearity. There are 144 million parameters in the network and the CNN features are vectors of 4096 dimensions.

For each object category, we assign a class label which is a vector of N dimensions (N is the total number of object classes). If an image belongs to the ith class, the ith element of its label vector is set to 1, and all the other elements are set to 0.

The extracted features are fed into the pattern association network as conditioned stimulus and the class label is input as unconditioned stimulus.

6.3.2 Increment learning

Instead of the simple Hebbian learning rule in the original pattern association model, the Leabra learning framework is introduced into our model for pattern learning. Fig. 6.3 summarizes the fundamental mechanism of the Leabra learning framework, which learns patterns in an error-driven and associative, biologically feasible way, and represents a balance between low-level, biological models and more abstract, computational models[30]. The Leabra learning framework uses a combination of a pure Hebbian learning and error-driven learning which amounts to the Contrastive Hebbian Learning (CHL)[31]. The learning process consists of two activation stages: a minus stage and a plus stage. The minus stage corresponds to the expected output activation, while the plus stage corresponds to the target output activation[32].

The CHL weight update resulting from presynaptic firing x_j and postsynaptic firing y_i can be written as:

$$\Delta CHL_{ij} = x_j^+ y_i^+ - x_j^- y_i^- \tag{6.4}$$

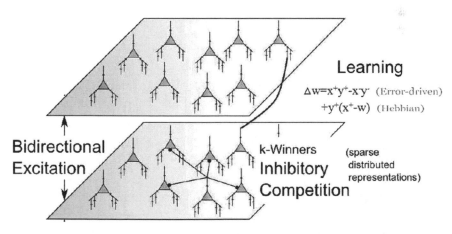

FIGURE 6.3   Summary of the learning mechanisms in the Leabra framework.

The + and − superscripts denote the plus and minus stages respectively. The Hebbian learning rule can be described as:

$$\Delta Hebb_{ij} = y_i^+ (x_j^+ - w_{ij}) \tag{6.5}$$

Then the two learning rules are combined together proportionally by a weight parameter λ and form the following rule:

$$\Delta w_{ij} = \alpha \left[\lambda \Delta_{Hebb} + (1 - \lambda) \Delta_{CHL} \right] \tag{6.6}$$

where $\lambda \in (0, 1)$ is a weight coefficient which adjusts the proportion between Hebbian learning and CHL learning, and α is the learning rate.

We use a k-winners-take-all (kWTA) function[7] to achieve sparse distributed representation. We take the top k (10% of the neurons) active units calculated by the following inhibitory function:

$$y_i = \begin{cases} y_i & if\ y_i \geq \theta \\ 0 & otherwise \end{cases} \tag{6.7}$$

where θ is the inhibitory threshold which allows the top k units to keep active, which can be achieved by sorting all the neural activations and take the value of the kth most excited units.

As illustrated in Fig. 6.2, in our model, the unconditioned stimulus corresponds to the class label in which only one element is non-zero. As a result, for the ith class, only the ith column of weight matrix w is non-zero and all the other columns are 0. So in the final weight matrix w, the ith column corresponds to the weight of the ith class. Learning the ith category only involves updating the ith column and leaves other columns unchanged. Thus, catastrophic forgetting would not occur in our model because there is no interference among the weight of different categories. Similarly, learning a new category only involves inserting a new column into the synaptic matrix. In this way, incremental learning can be easily achieved.

6.3.3 Recall

When the feature vector of a test image is presented, the activation h_i of neuron i is the normalized dot product between the ith column of synapse matrix w and the feature vector x, which can be described as:

$$h_i = \frac{\sum_{j=1}^{D} x_j w_{ij}}{\|x\|_2 \|w_{\cdot i}\|_2} \tag{6.8}$$

where $\|x\|_2$ denote the L2-norm of vector x and $w_{\cdot i}$ represents the ith column of matrix w.

After calculating the activation of all neurons, we seek for the neuron which has the maximum activation:

$$i* = \arg\max \{h_i\} \tag{6.9}$$

If the activation of the i^*th neuron exceeds a preset threshold Θ, then the image is considered to be studied already and it will be classified to the i^*th class. Otherwise, it will be considered as a new pattern which has not been studied before.

$$result = \begin{cases} \text{new pattern} & h_{i_*} < \Theta \\ i*th \text{ class} & otherwise \end{cases} \tag{6.10}$$

6.4 EXPERIMENTAL RESULTS

To evaluate the performance of the proposed model, we perform experiments on three benchmark datasets (Caltech5, Caltech256, and Scene15). The proposed model (IPAMM) is compared with some prevalent neural models associated with memory modelling, including Self-supervised ART[33], Biased ART[34], Self-organizing Incremental Neural Network (SOINN)[35], BP Neural Network (BPNN)[36], and Bayesian MLP network[37]. Meanwhile, the proposed model is also compared with Support Vector Machine (SVM)[38]. Experiments are conducted on a laptop computer (CPU: Intel CoreTM i5-4200M @ 2.50GHZ, RAM: 4G; OS: Win7). In our model, the learning rate α is set to $1/M$ (M is the number of training samples), and the learning rate of Bayesian MLP network is set to 0.01. Other parameters of BPNN and Bayesian MLP network are presented in Table 6.2.

TABLE 6.2 Parameters in BPNN and MLP network

Hidden units	Epochs	Mean Squared Error	Gradient
10	100	1×10^{-7}	1×10^{-7}

6.4.1 Caltech5 dataset

Caltech5 dataset includes five object categories (airplanes, rear-cars, faces, leaves, and motorcycles), which are available at http://www.robots.ox.ac.uk/~vgg/data3.html. In this experiment, we select various training samples (5, 15, 30, 45, and 60) from each category, and the rest are used for testing. Image features are extracted based on the three aforementioned methods: ScSPM, Sparse-HMAX, and CNN.

To select an appropriate value for the weight parameter λ in Equations (6.5) and (6.6), we run the proposed model on Caltech5 dataset under varying number of training samples and different values of λ, and present the results in Table 6.3. Experimental results show that the proportion of error-driven learning increases accordingly as the number of training samples increases, which is consistent with the fact that error-driven learning is good at dealing with longer-term, larger-scale problems. As illustrated in Table 6.3, when there are fewer training images, Hebbian learning plays a dominant role and better performance can be obtained with $\lambda = 0.9$. However, as the number of training images increases, better performance can be achieved with $\lambda = 0.6$ or 0.7. So in the following experiments, λ is set to 0.9 when there are small number of training samples (no more than 40); and λ is set to 0.6 under the condition of large number of training samples.

TABLE 6.3 Classification results under different number of training samples and different values of λ

Number of Training	$\lambda = 0.1$	$\lambda = 0.2$	$\lambda = 0.3$	$\lambda = 0.4$	$\lambda = 0.5$	$\lambda = 0.6$	$\lambda = 0.7$	$\lambda = 0.8$	$\lambda = 0.9$
5	0.822	0.877	0.905	0.926	0.944	0.955	0.965	0.968	**0.971**
10	0.792	0.903	0.938	0.960	0.976	0.982	**0.987**	**0.987**	**0.987**
15	0.905	0.956	0.977	0.984	0.985	0.986	0.987	**0.989**	**0.989**
20	0.965	0.971	0.976	0.978	0.982	**0.984**	0.983	**0.984**	**0.984**
25	0.968	0.971	0.970	0.970	0.974	0.984	0.984	0.984	**0.985**
30	0.92	0.959	0.98	0.985	0.984	0.985	0.986	**0.987**	**0.987**
35	0.907	0.947	0.964	0.977	**0.983**	**0.983**	**0.983**	**0.983**	**0.983**
40	0.905	0.947	0.959	0.965	0.979	0.979	0.979	0.979	**0.980**
45	0.915	0.977	**0.986**	**0.986**	**0.986**	**0.986**	0.983	0.982	0.982
50	0.930	0.967	0.967	0.971	0.976	**0.984**	**0.984**	**0.984**	**0.984**
55	0.946	0.967	0.971	0.978	0.985	0.985	**0.986**	0.985	0.985
60	0.917	0.975	0.99	0.991	**0.993**	**0.993**	**0.993**	0.991	0.991

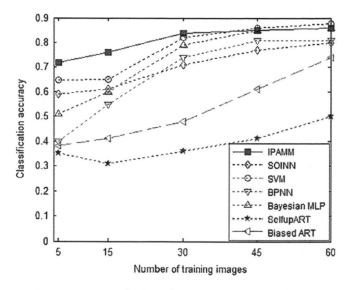

FIGURE 6.4 Comparison results based on Sparse-HMAM features.

Fig. 6.4 gives the comparison results of IPAMM and the contrast models including SOINN, BPNN, SVM, Bayesian MLP, Self-supervised ART (denoted as SelfupART), and Biased ART based on Sparse-HMAM features. In general, the more training samples there are, the higher accuracy will be achieved for all the models. IPAMM has significantly outperformed all contrast models except SVM in this experiment. When the number of training images is less than 45, IPAMM shows better performance than SVM. But when there are more than 45 training samples, our model performs a little worse than SVM. The proposed model shows a great advantage when there are a small number of training samples.

Fig. 6.5 presents the classification performance of IPAMM and the contrast models based on CNN features. Similarly, our method shows outstanding performance with small number of training samples. When there are a large number of training samples, IPAMM has comparative performance with SVM and self-supervised ART. Generally speaking, our model demonstrates the best performance in this experiment.

Table 6.4 presents the processing speed of IPAMM and the contrast methods under CNN features. The processing time is the total time (second) spent on both training and recognition stages. As the number of training images increases, the processing time will increase

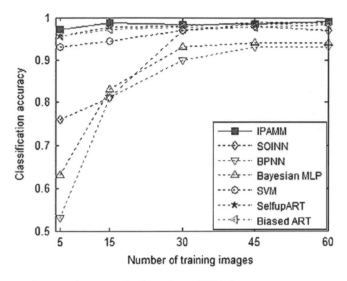

FIGURE 6.5 Comparison results based on CNN features.

TABLE 6.4 Processing speed (second) of IPAMM and contrast models under varying training samples

Methods	5	15	30	45	60
IPAMM	0.514	0.526	0.492	0.493	0.479
SOINN	0.487	0.598	0.967	1.121	1.552
SVM	0.281	0.397	0.691	0.897	1.046
SelfupART	0.273	0.265	0.274	0.270	0.285
Biased ART	1.143	1.155	1.274	1.243	1.327
BPNN	168	210	293	314	625
Bayesian MLP	588	653	700	725	766

accordingly for most models (SOINN, SVM, BPNN, and Bayesian MLP). However, as the training stage of our model is not very time-consuming, so the total processing time of our model changes little as the number of training samples varies. Though SVM shows better time performance than our model with fewer training samples; however, when there are more training samples, IPAMM shows advantage over SVM in time efficiency. Though Self-supervised ART shows the best time performance in this experiment, its classification accuracy is lower than that of our method.

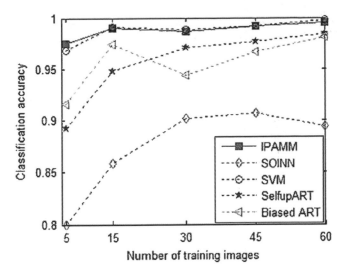

FIGURE 6.6   Comparison results based on ScSPM features.

Fig. 6.6 presents the comparison among IPAMM and contrast models under ScSPM features. Because of the limited hardware configuration, our computer fails to run BPNN and Bayesian MLP network with high-dimensional (21,504) features, so the two models are left out in this experiment. It is notable that the proposed model continues to significantly outperform SOINN, Self-supervised ART, and Biased ART, and shows equivalent performance with SVM.

It can be seen that the proposed model shows outstanding performance despite various image representations and different number of training samples used in this experiment. A prominent advantage of our method is its outstanding classification performance with small number of training samples, and it greatly outperforms all contrast models. IPAMM can achieve satisfactory results with only 15 training images and is more appropriate for the case of small training set. In the case of large number of training samples (larger than 45), our method has also outperformed the contrast models except SVM which shows comparable performance. Though in Fig. 6.4, our model performs a little worse than SVM, it does not affect overall performance. Another advantage of IPAMM model is high time efficiency. Due to its simple architecture, IPAMM achieves better time performance than the contrast models except Self-supervised ART. However, our model steadily outperforms Self-supervised ART in classification accuracy. In summary, our model demonstrates excellent classification performance steadily under various image presentation methods and has high efficiency as well.

6.4.2 Caltech256 dataset

In this experiment, we select three object categories (bicycle, treadmill, and revolver) from Caltech256 dataset which is available at http://www.vision.caltech.edu/Image_Datasets/ Caltech256/. For each category, half of the images are used for training and the remaining images are used for testing. As mentioned previously, Sparse-HMAX, ScSPM, and CNN are used to extract features from these images. From the results presented in Table 6.5, we can find that IPAMM shows the best performance under Sparse-HMAX and ScSPM features. But for CNN features, the classification accuracy of IPAMM is a little lower than ART methods. Relatively speaking, ART methods only do well in CNN features, while IPAMM has better versatility and can be well applied to various image representations.

6.4.3 Scene15 dataset

To evaluate our model's performance on scene recognition task, we perform experiment on scene15 dataset which is available at http://www-cvr.ai.uiuc.edu/ponce_grp/data/. First image features are extracted based on ScSPM algorithm, and then fed into IPAMM, SOINN, SVM, and Self-supervised ART to perform classification tasks. Experimental results (Fig. 6.7) show that the proposed model demonstrates an advantage over all the comparison methods though there are more categories in this experiment.

TABLE 6.5   Comparison results under Sparse-HMAX, ScSPM, and CNN features

Methods	Sparse-HMAX	ScSPM	CNN
IPAMM	**0.97**	**0.96**	0.99
SOINN	0.95	0.85	0.99
SVM	0.93	0.95	0.99
Selfup ART	0.84	0.95	**1.00**
Biased ART	0.61	0.75	**1.00**
BPNN	0.87	\	0.96
Bayesian MLP	0.92	\	0.77

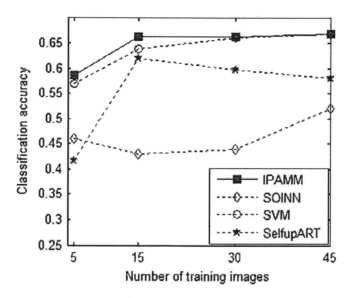

FIGURE 6.7   Comparison results on Scene dataset under ScSPM features.

6.5 DISCUSSION

6.5.1 Capacity

Storage capacity is an important performance index for associative memories. Our model can learn patterns incrementally, and each category corresponds to an individual column of synapse matrix. So learning a category only involves updating a single column and would not affect the categories which have already been studied. In addition, adding new categories need not learn the old categories again. As a result, our model does not encounter the problem of inadequate storage capacity which has occurred in some memory models such as Complementary Learning Systems (CLS)[39].

6.5.2 Speed

Many neural networks encounter a challenging computational burden and many attempts[40–42] have been made to reduce computational cost and improve the processing speed. Due to the simple learning mechanism and network structure, the proposed model has achieved high time efficiency as demonstrated in the previous section (Table 6.4).

As our model only contains one learning layer, learning is very fast. A pair of conditioned stimulus and unconditioned stimulus makes the association be learned, and the learning process need not be repeated. Assume the number of training samples is M, the dimension of feature vector is D and the number of object classes is N, then the time complexity of pattern learning is $O(MDN)$, which means that the processing time is linearly proportional to the number of training samples M, feature dimension D, and the number of classes N.

Recall is also fast in the proposed model, once the conditioned stimulus input x_j is presented, it would be synchronously applied to the synapse w_{ij}, and then the activation h_i can be computed immediately. Meanwhile, the neuron with the highest activation can also be found instantaneously.

6.5.3 Future work

The proposed model shows excellent performance and outperforms all contrast models under the condition of small number of training samples. However, when there are large numbers of training samples (especially larger than 45), our model does not demonstrate the best performance in all tasks. Though combining error-driven learning can improve the performance in some degree, further work still needs to be done on the learning mechanism to improve the recognition performance especially when there are a large number of training samples. Meanwhile, our model includes only one learning layer; though a simple learning structure can achieve better time efficiency, adding additional hidden layers may further improve the performance of the model. In addition, the value of parameter λ in Equation (6.6) is estimated by limited experiments in a very simple way. More optimal value may be obtained by learning, and the classification performance could be further improved accordingly.

6.6 CONCLUSIONS

In this chapter, we present an increment pattern association memory model (IPAMM) to realize increment learning and rapid retrieval of visual information. The Leabra learning mechanism, which combines Hebbian learning with error-driven learning, is introduced into our model instead of the pure Hebbian learning rule. Meanwhile, increment learning is achieved by assigning individual synapse weight to different patterns. We have conducted a serial of experiments on benchmark datasets based on three feature-extraction methods and compared

IPAMM with some prevalent models. Though the proposed model does not achieve the best performance in all tasks, it has significantly outperformed most of neural models such as SOINN, BPNN, Bayesian MLP, and biased ART both in time efficiency and classification accuracy, and shows comparable performance with SVM. The prominent advantage of IPAMM is its outstanding classification performance with small training set and high time efficiency, which can well satisfy the practical engineering applications. Meanwhile, the proposed model is able to learn patterns incrementally, and learning a new pattern only involves inserting a new column to the synaptic matrix without destroying the stored patterns.

REFERENCES

1. Fei-Fei, L., and Perona, P. 2005, June. A bayesian hierarchical model for learning natural scene categories. *In 2005 IEEE Computer Society Conference on Computer Vision and Pattern Recognition (CVPR'05)* 2: 524–531.
2. Cisek, P., Drew, T., and Kalaska, J. (Eds.). 2007. *Computational Neuroscience: Theoretical Insights into Brain Function.* Elsevier.
3. Tao, D., Li, X., Wu, X., and Maybank, S. J. 2008. Geometric mean for subspace selection. *IEEE Transactions on Pattern Analysis and Machine Intelligence* 31(2):260–274.
4. Albus, J. S. 2010. A model of computation and representation in the brain. *Information Sciences* 180(9):1519–1554.
5. Kirkpatrick, J., Pascanu, R., Rabinowitz, N., Veness, J., Desjardins, G., Rusu, A. A., and Hadsell, R. 2017. Overcoming catastrophic forgetting in neural networks. *Proc Natl Acad Sci USA* 114 (13):3521–3526.
6. Dean, T. 2005. A computational model of the cerebral cortex. *In: National Conference on Artificial Intelligence.* AAAI Press:938–943.
7. Norman, K. A., and O'Reilly, R. C. 2003. Modelling hippocampal and neocortical contributions to recognition memory: a complementary-learning-systems approach. *Psychological Review* 110(4):611–646.
8. Hattori, M. 2014. A biologically inspired dual-network memory model for reduction of catastrophic forgetting. *Neurocomputing* 134:262–268.
9. Sakurai, N., HattorI, M., and Ito, H. 2002. SOM associative memory for temporal sequences. *International Joint Conference on Neural Networks,* vol 1. IEEE, pp. 950–955.
10. Grossberg, S. 2013. Adaptive resonance theory: how a brain learns to consciously attend, learn, and recognize a changing world. *Neural Networks the Official Journal of the International Neural Network Society* 37(1):1–47.

11. Yu, J., Rui, Y., and Chen, B. 2014. Exploiting click constraints and multi-view features for image re-ranking. *IEEE Transactions on Multimedia* 16(1):159–168.

12. Yu, J., Tao, D., Wang, M., Rui, Y. 2015. Learning to rank using user clicks and visual features for image retrieval. *IEEE Transactions on Cybernetics* 45(4):767–779.

13. Yu, J., Rui, Y., Tao, D. 2014. Click prediction for web image reranking using multimodal sparse coding. *IEEE Transactions on Image Processing* 23(5): 2019–2032.

14. Hong, C., Yu, J., Tao, D., Wang, M. 2015. Image-based three-dimensional human pose recovery by multiview locality-sensitive sparse retrieval. *IEEE Transactions on Industrial Electronics* 62(6):3742–3751.

15. Su, Y. 2018. Robust video face recognition under pose variation. *Neural Processing Letters* 47(1): 277–291.

16. Liu, W., Tao, D., Cheng, J., Tang, Y. 2013. Multiview Hessian discriminative sparse coding for image annotation. *Computer Vision & Image Understanding* 118(1):50–60.

17. Tao, D., Li, X., Wu, X., Maybank. S. J. 2007. General tensor discriminant analysis and Gabor features for gait recognition. *IEEE Transactions on Pattern Analysis & Machine Intelligence* 29(10):1700–1715.

18. Yu, J., Zhang, B., Kuang, Z., Lin, D., Fan, J. 2017. Iprivacy: image privacy protection by identifying sensitive objects via deep multi-task learning. *IEEE Transactions on Information Forensics & Security* 12(5):1005–1016.

19. Yu, J., Yang, X., Gao, F., Tao, D. 2016.Deep multimodal distance metric learning using click constraints for image ranking. *IEEE Transactions on Cybernetics PP* (99):1–11.

20. Hong, C., Yu, J., Wan, J., Tao, D., Wang, M. 2015. Multimodal deep autoencoder for human pose recovery. *IEEE Transactions on Image Processing* 24(12):5659–5670.

21. Palm, G. 2013. Neural associative memories and sparse coding. *Neural Networks* 37:165–171.

22. Rolls, E. T. 2016. *Cerebral cortex: principles of operation*. Oxford: Oxford University Press.

23. Rolls, E. T., Treves, A. 1998. *Neural networks and brain function*. Oxford: Oxford University Press.

24. Deng, L., Gao, M., Wang, Y. 2018. Increment learning and rapid retrieval of visual information based on pattern association memory. *Neural Processing Letters* 48(3):1597–1610.

25. O'Reilly, R. C., Munakata, Y., Frank, M. J., Hazy, T. E. 2012. *Computational Cognitive Neuroscience*. Wiki Book, 1st Edition.

26. Yang, J., Yu, K., Gong, Y., Huang, T. 2009. Linear spatial pyramid matching using sparse coding for image classification. *In: IEEE Conference on Computer Vision and Pattern Recognition (CVPR 2009)*. IEEE, pp. 1794–1801.

27. Wang, Y., Deng, L., 2016. Modelling object recognition in visual cortex using multiple firing k-means and non-negative sparse coding. *Signal Processing* 124:198–209.

28. Simonyan, K., Zisserman, A. 2014. Very deep convolutional networks for large-scale image recognition. *arXiv preprint arXiv*:1409.1556.

29. Hebb, D. O. 1949. *The organization of behavior: a neuropsychological theory*. New York: Wiley.

30. Liu, W., Zha, Z. J., Wang, Y., Lu, K., Tao, D. 2016. P-Laplacian regularized sparse coding for human activity recognition. *IEEE Transactions on Industrial Electronics* 63(8):5120–5129

31. O'Reilly, R. C., Munakata, Y. 2000. *Computational explorations in cognitive neuroscience: Understanding the mind by simulating the brain*. Cambridge: MIT Press.

32. Ketz, N., Morkonda, S. G., O'Reilly, R. C. 2013. Theta coordinated error-driven learning in the hippocampus. *PLoS Comput Biol* 9(6):e1003067.

33. Amis, G., Carpenter, G. 2009. Self-supervised ARTMAP. *Neural Networks* 23:265–282.

34. Carpenter, G. A., Gaddam, S. C. 2010. Biased ART: a neural architecture that shifts attention toward previously disregarded features following an incorrect prediction. *Neural Networks* 23:435–451.

35. Shen, F., Hasegawa, O. 2006a. An incremental network for on-line unsupervised classification and topology learning. *Neural Networks* 19:90–106.

36. Sadeghi, B. H. M. 2000. A BP-neural network predictor model for plastic injection molding process. *Journal of Materials Processing Technology* 103(3): 411–416.

37. Vehtari, A., Lampinen, J. 2000. Bayesian MLP neural networks for image analysis. *Pattern Recognition Letters* 21(13–14): 1183–1191.

38. Chang, C., Lin, C. 2011. LIB-SVM: a library for support vector machines. ACM *Trans. Intell. Syst. Technol.* 2(3): 1–27.

39. Norman, K. A. 2010. How hippocampus and cortex contribute to recognition memory: revisiting the complementary learning systems model. *Hippocampus* 20(11):1217–1227.

40. Vanhoucke, V., Senior, A., Mao, M. Z. 2011. Improving the speed of neural networks on CPUs. *In: Proc. Deep Learning and Unsupervised Feature Learning NIPS Workshop*, vol 1, p.1–4.

41. Werner, G. Á., Hanka, L. 2016. Tuning an artificial neural network to increase the efficiency of a fingerprint matching algorithm. *In: 2016 IEEE 14th International Symposium on Applied Machine Intelligence and Informatics (SAMI)*. IEEE, pp. 105–109.

42. Han, S., Liu, X., Mao, H., et al (2016) Eie: efficient inference engine on compressed deep neural network. *Acm Sigarch Computer Architecture News* 44(3):243–254.

APPENDIX

```
%% Programmed by Matlab
```

1. PatternAssociationDemo.m

```
% Pattern association network demonstration program
% The background to these networks is provided in
% Rolls ET (2016) Cerebral Cortex: Principles of
Operation (Oxford University Press) Appendix 2 (B)
    http://www.oxcns.org
% which is available at
by following the link to the book.
% The instructions and exercises for this code are in
% Rolls ET (2016) Cerebral Cortex: Principles of
Operation (Oxford University Press) Appendix 4 (D)
    http://www.oxcns.org
% which is available at
by following the link to the book.
% The background to these networks is also avail-
able in
% Rolls ET (2008) Memory, % Attention and Decision-
Making (Oxford University Press) Appendix 2, which is
available at
    http://www.oxcns.org/papers/RollsMemoryAttentionAndDecision
MakingContents+Appendices1+2.pdf
%

clear;
clc;
%%load dataset , including XTrain , XTest , TrainLabel
and TestLabel
load datasets\caltech5CNN40.mat;
%%
nTypes=max(TrainLabel);
trainLength=length(TrainLabel);
US=zeros(nTypes,trainLength);
num=zeros(nTypes);
for i=1:trainLength
```

```
  US(TrainLabel(i),i)=1;
  num(TrainLabel(i))= num(TrainLabel(i))+1;
end
testLength=length(TestLabel);
XTestL=zeros(nTypes,testLength);
for i=1:testLength
  XTestL(TestLabel(i),i)=1;
end
% Make some training patterns
TrainPatts =XTrain;
TrainPatts=mapstd(TrainPatts);
TestPattern = XTest;
TestPattern=mapstd(TestPattern);

N =size(US,1);      % number of neurons in the pattern
associator
nSyn = size(TrainPatts,1);    % number of synapses on
each neuron
% parameters for some training patterns that overlap
by shift
InputSparseness = 0.25; % the sparseness of the input
CS patterns is the number of bits on for any one
pattern
Learnrate = 1 / trainLength; % Not of especial sig-
nificance in an autoassociation network, but this
keeps the weights within a range
% Learnrate = 1 / 30; % Not of especial significance in
an autoassociation network, but this keeps the
weights within a range
OutputSparseness = 1 / N; % the US patterns have one
neuron on in this simple simulation

Actvn = zeros(1, N); % activation
Rate = zeros(1, N);  % output firing rates
nepochs = 5;      % the number of times that the network
is allowed to update during testng
```

```
% ..........Train.............. See Rolls 2008
B.2.7.1 for why an associative learning rule with
heterosynaptic long term depression is useful
startPoint=1;
for type =1:nTypes
  SynMat= zeros(nSyn, N);
  endPoint=startPoint+num(type)-1;
  for patt=startPoint:endPoint
    for neuron = 1 : N
      for syn = 1 : nSyn
        preSynRate = TrainPatts(syn, patt); % the CS
patterns used for training
        postSynRate = US(neuron, patt);
          weight_change =Learnrate * postSynRate *
preSynRate; % an associative, Hebbian, synaptic
modification rule

        SynMat(syn, neuron) = SynMat(syn, neuron) +
weight_change;
      end
    end
  end
  cell{type,1}=SynMat;
  startPoint=endPoint+1;
  % pause
end



% ........... Test........................
disp('Testing');
R=zeros(1,nTypes);
Rate1=zeros(1,nTypes);
nCorrect = 0;     % the number of correctly recalled
patterns
predictLabel=zeros(testLength,1);
nPatts=testLength;
Actvns=zeros(nTypes,N);
```

```
for patt = 1 :nPatts
% calculate activation
maxValue=-1000;
maxIndex=0;
  for n=1:nTypes
   for neuron = 1 : N
       PreSynInput = TestPattern(:, patt); % OR the
distorted retrieval cue to test for generalization
           Actvn(neuron) = WindowedPatchDistance
(PreSynInput,cell{n,1}(:, neuron));
     Actvns(n,:)=Actvn;
    end
    maxtmp=max(Actvn);
    if maxValue<maxtmp
      maxValue=maxtmp;
      maxIndex=n;
    end
    Rate1(n)=Actvns(n,n);
  end %%end for n

  scale = 1.0 / (max(Rate1) - min(Rate1));

  Rate = (Rate1 - min(Rate1)) * scale;        % scale
Activation to 0-1.

  % Now convert the activation into a firing rate using a
binary or linear threshold activation functon
  % The threshold is selected in this case artificially
by sorting
  % the array to determine where the threshold should
be set to achieve the required sparseness
  % In the brain, inhibitory neurons are important in
setting the proportion of excitatory neurons that are
active
  tmp = sort(Rate, 'descend');
  el = floor(length(Rate1) * OutputSparseness);%
  if el < 1
    el = 1;
```

```
  end
  Threshold = tmp(el);

   for neuron = 1 : length(Rate)
      if Rate(neuron) >=Threshold % threshold binary
activation function
          Rate(neuron) = 1; % comment out this line for a
threshold linear activation function
      else
        Rate(neuron) = 0;
      end
   end

  % check performance
    R = corr(XTestL(:,patt), Rate'); % correlation
coefficient
  disp(['Patt ', num2str(patt), ' Correlation of re-
called firing rate vector with training pattern = ',
num2str(R)]);
%    if r > 0.98 % the criterion of correct is a corre-
lation of 0.98 between a recalled pattern and the
stored pattern
    if R > 0.98 % the criterion of correct is a correla-
tion of 0.98 between a recalled pattern and the stored
pattern
          predictLabel(patt,1)=maxIndex;
          disp(['It belongs to ',num2str(maxIndex),'
catagory']);
    else
      disp(('Tt fails to recall '));
    end

end %%end for patt
PercentCorrect = nCorrect / nPatts * 100;
disp(['Percentage  correct  recall  =  ',  num2str
(PercentCorrect)]);
successrate = mean(TestLabel==predictLabel);
```

```
disp(['Percentage correct recognition = ', num2str
(successrate)]);
```

2. ClassificationDemo.m

```
% evaluate the performance of the IAPMM model
addpath ('osu-svm') %put your own path to osusvm here
addpath('libsvm');
addpath('sfam');
addpath('FuzzyARTMAP');
addpath('SelfSupARTMAP');
addpath('bARTMAPcode');
addpath('soinn');
clear;
clc;

load datasets\256_ObjectCategories_40.mat;

XTrain=double(XTrain);
XTest=double(XTest);
XTrain=mapstd(XTrain);
XTest=mapstd(XTest);
% %% osu-svm classifier
tic
disp('osu-svm:');
Model = HCLSosusvm(XTrain,TrainLabel); %training
[ry,rw] = HCLSosusvmC(XTest,Model); %proubleedicting
new labels
successrate = mean(TestLabel==ry) %a simple
classification scoreModel =
toc
%% libsvm model
tic
disp('libsvm:');
XTrain=zscore(XTrain);
XTest=zscore(XTest);
XTrain=XTrain';
XTest=XTest';
```

```
[bestCVaccuracy,bestc,bestg]=SVMcgForClass
(TrainLabel,XTrain');
%model = svmtrain(TrainLabel,XTrain','-s 0 -t 2 -c
3.03 -g 0.0039');
model = svmtrain(TrainLabel,XTrain','-s 0 -t 2 -c
3.03 -g 0.0068');
[predictlabel,accuracy, dec_values] = svmpredict
(TestLabel,XTest',model);
model = svmtrain(TrainLabel,XTrain');
[predictlabel,accuracy, dec_values] = svmpredict
(TestLabel,XTest',model);
toc

% % % SFAM classifer
tic
disp('SFAM:');
  % create network
XTrain1=mapminmax(XTrain', 0, 1);
XTest1=mapminmax(XTest', 0, 1);
XTrain1=XTrain';
XTest1=XTest';
% %   XTest= mapminmax('apply',XTest,ps)
net = create_network(size(XTrain1,2));
%   % change some parameters as you wish
net.epochs = 50;
%   % train the network
tnet = train(XTrain1, TrainLabel, net, 100);
%   % test the network on the testdata
r = classify(XTest1, tnet, TestLabel, 50);
toc

%  %% ARTMAP classifer
%  %%
%  % The datastruct fields are:
%  % training_input: [f features X m records]
%  % training_output: [m labels X 1]
%  % test_input: [f features X n records]
%  % test_output: [n labels X 1]
```

```
% % description: 'dataset_title'
% % descriptionVerbose: 'A more verbose description
of the dataset'
tic
%
training_input=XTrain;
training_output=TrainLabel;
test_input=XTest;
test_output=TestLabel;
description='coil20';
descriptionVerbose='coil20';
%
% save coil20data training_input training_output
test_input test_output description description
Verbose;
%
disp('ARTMAP:');
dataStruct=load('coil20data.mat');
% % [a,b,c] = fuzzyARTMAPTester(dataStruct);
% % successrate2=mean(dataStruct.test_output==a)
%
[a1,b1,c1] = biasedARTMAPTester(dataStruct,200);
successrate3=mean(dataStruct.test_output==a1)
%
toc


%% SelfSupARTMAP classifier
tic
disp('SelfSupARTMAP:');
% XTrain2=mapminmax(XTrain, 0, 1);
% XTest2=mapminmax(XTest, 0, 1);
XTrain2=XTrain;
XTest2=XTest;
am = SelfSupAM_trainStage1(XTrain2,TrainLabel',
'E2',1);
acc1 = 100*mean(SelfSupAM_test(am,XTest2)==
TestLabel');
```

```
disp(sprintf('Accuracy after Stage 1 training: %1.1f
%%',acc1));
am.SelfSupAM_trainStage2(XTrain2);
acc2 = 100*mean(SelfSupAM_test(am,XTest2)==
TestLabel');
disp(sprintf('Accuracy after Stage 2 training: %1.1f
%%',acc2));
toc

%% SOINN classfier
tic
disp('SOINN:');
XTrain=mapstd(XTrain);
XTest=mapstd(XTest);

nTypes=max(TrainLabel);
trainLength=length(TrainLabel);
num=zeros(nTypes);
for i=1:trainLength
  num(TrainLabel(i))= num(TrainLabel(i))+1;
end
% Make some training patterns
TrainPatts =XTrain;
% TrainPatts=mapstd(TrainPatts);
% load teapest.mat;
testLength=length(TestLabel);
TestPattern = XTest;
% TestPattern=mapstd(TestPattern);

%%training SOINN
startPoint=1;
for ii =1:nTypes
  maxpool = [];
  endPoint=startPoint+num(ii)-1;
  for patt=startPoint:endPoint
   maxpool=[maxpool TrainPatts(:, patt)];
  end
  [node1,threshold1]=soinn_asc(maxpool',50,50);
```

```
  node3{ii} =node1;
  thre3{ii}=threshold1;
  startPoint=endPoint+1;
  % pause
end

% test
studmun=0;
recallmun=0;
predictLabel=zeros(testLength,1);
for patt = 1 :testLength
 xx- TestPattern(:, patt);
 xx=xx';
 for m=1:nTypes
   node2=node3{m};
   thre2=thre3{m};
     dispre=sqrt(sum((repmat(xx,size(node2,1),1)-
node2).^2'));
   [valuemin indexmin]=min(dispre);
   amin(m,:)=[valuemin indexmin];
   node(m,:)=node2(indexmin,:);
   thremin(m)=thre2(indexmin);
   thremax(m)=max(thre2);
 end
 [c,d]=min(amin(:,1));
%%
if c>thremax(d)
%   disp(['Patt ', num2str(patt),'is not studied']);
else
   studmun=studmun+1;
   predictLabel(patt,1)=d;
%         disp(['Patt ', num2str(patt),' belongs to
',num2str(d),' catagory']);
end

end
studmun=studmun/testLength*100;
```

```
disp(['The studmun rate is ',num2str(studmun)]);

successrate = mean(TestLabel==predictLabel);
disp(['Percentage correct recognition = ', num2str
(successrate)]);
toc
%}
```

Memory modelling based on free energy theory and the restricted Boltzmann machine

7.1 INTRODUCTION

Friston believes that cortical response can be considered as a process in which the brain tries to minimize the free energy caused by stimulation and encode the factors that induce the stimulus. Similarly, learning results from changes in synaptic efficacy that minimizes free energy, and the process of minimizing free energy corresponds to the process of remembering things[1]. The theory of free energy minimization provides a theoretical explanation of brain perceptual reasoning. The minimum value of energy function corresponds to the most stable state of the system. The more orderly the system or the more concentrated the probability distribution, the smaller the energy of the system. We can use Bayesian model to simulate the process of brain perception and reasoning. Probabilistic generative model can explain many aspects of

DOI: 10.1201/9781003281641-7

anatomy and physiology of the brain system, such as hierarchical structure of cortex, network structure of forward and feedback connection and functional asymmetry of connection[2]. In terms of synaptic physiology, it predicted the associated plasticity and the firing timing-dependent plasticity of dynamic models. It explains the long-term effect of the receptive field on the physiology. In psychophysics, it explains the behavioural relevance of these physiological phenomena, such as priming effect and global priority. Obviously, both perceptual reasoning and learning depend on the minimization of free energy or the inhibition of prediction errors. The brain free energy theory can explain the intimate relationship between perception and behaviour[1].

Friston's free energy theory provides a powerful theoretical framework to describe how the sensory cortex extracts information from noisy stimuli. However, this framework is highly conceptual and theoretical[3], and there is no successful application result yet. RBM is an energy-based model derived from statistical mechanics and is a powerful probabilistic generative model. Based on this, this chapter combines Friston's free energy theory with RBM to realize the learning and memory of visual information.

7.2 THEORY OF FREE ENERGY

Friston et al.[4] considered that the free energy is a scalar function of the global density and the current sensory input, which consists of two parts:

$$
\begin{aligned}
F &= -\int q(v)\ln \frac{p(y, v)}{q(v)} dv \\
&= -<\ln p(y, v)>_q + <\ln q(v)>_q
\end{aligned}
\tag{7.1}
$$

where $<\cdot>_q$ represents the expectation under density q, $p(y, v)$ represents the generation density, and a generative model is defined to generate the joint probability distribution of input sample y and cause v. $q(v)$ represents the overall density and is used to approximately identify the density $p(v|y)$. The first term of Equation (7.1) represents the energy expectation of the system under the overall density, and the second term is the negative entropy of the overall density. Thus, it can be seen that the free energy defines the overall density of the generation model of a system and the cause of a model.

The generative model $p(y,v)$ can be used to associate cause v and parameter θ with input y for parameter learning and cause inference:

$$p(y, v; \theta) = p(y|v; \theta)p(v; \theta) \tag{7.2}$$

Given input y, the identification (conditional) density of its cause v can be obtained from the following equation:

$$p(v|y; \theta) = \frac{p(y|v; \theta)p(v; \theta)}{p(y; \theta)} \tag{7.3}$$

It is very difficult to invert the density directly from the generated model, so an approximate identification density $q(v)$ is used which is consistent with the generated model. The expected estimation of the density corresponds to inference, and the parameters of the estimated generated model correspond to learning, which can be mapped directly to the two steps of the expected maximization (EM) algorithm (Step E and Step M). Assuming that the approximate recognition density satisfies the Gaussian distribution, the whole process can be divided into two stages: learning (M) stage and inferring (E) stage. The learning stage is used to learn the corresponding model parameters, and the inference stage is used to estimate the recognition density[5]. Both phases involve minimizing a function F of the above density, which corresponds to the free energy in physics:

$$F = -\ln p(y; \theta) + KL\{q(v), p(v|y; \theta)\} \tag{7.4}$$

The objective function F in Equation (7.4) contains two terms. The first term is the likelihood distribution of the input data under the generation model. The second term is the K-L divergence between the approximate recognition density and the recognition density, which is non-negative, which means that to minimize the objective function F (free energy), is to minimize the uncertainty of the input data. Free energy is a characteristic of thermodynamic systems. It represents the work required to keep the system in equilibrium, that is, when a system is in equilibrium, its free energy is minimal. For simplicity, the expectation of q(v) can be represented by ϕ. Then the inference stage is to reduce F according to the expected cause to ensure a good approximation of the identification density; in the learning stage, by changing the parameter, the generated model can better simulate the input distribution. That is,

$$\text{Inference phase (E): } \phi = min_\phi F$$

$$\text{Learning phase (M): } \theta = min_\theta F$$

Through the above two phases, an accurate and approximate maximum likelihood estimation can be obtained for the generation model describing the prior and generative distribution.

Under the Gauss assumption, the generation model generally has the following form[5]:

$$
\begin{aligned}
y &= g(v, \theta) + \varepsilon_y \\
v &= v_p + \varepsilon_p
\end{aligned}
\tag{7.5}
$$

where $Cov\{\varepsilon_y\} = \Sigma_y$ is the covariance of random variables in the generation process, and the prior of v is described by their expected v_p and covariance $Cov\{\varepsilon_p\} = \Sigma_p$. From formula (6.5), we can get:

$$
\begin{aligned}
L &= -\frac{1}{2}\xi_y^T \xi_y - \frac{1}{2}\xi_p^T \xi_p - \frac{1}{2}ln|\Sigma_u| - \frac{1}{2}ln|\Sigma_p| \\
\xi_y &= \Sigma_y^{-1/2}(y - g(\varphi, \theta)) \\
\xi_p &= \Sigma_p^{-1/2}(\varphi - v_p)
\end{aligned}
\tag{7.6}
$$

where L is the prediction error. If it is extended to the hierarchical model, then:

$$
\begin{aligned}
\xi_i &= \varphi_i - g_i(\varphi_{i+1}, \varphi_i) - \lambda_i \xi_i \\
&= (1 + \lambda_i)^{-1}(\varphi_i - g_i(\varphi_{i+1}, \theta_i))
\end{aligned}
\tag{7.7}
$$

where $\Sigma_i^{1/2} = 1 + \lambda_i$. In the neural network model, the prediction error is encoded by the activity unit marked as ξ_i. Then, the estimation of ϕ_i and parameters θ_i can be obtained by differentiating the objective function:

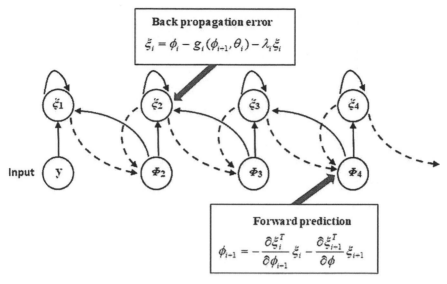

FIGURE 7.1   Schematic diagram of a hierarchical architecture[5].

Inference phase: $\varphi_{i+1} = \dfrac{\partial F}{\partial \varphi_{i+1}} = -\dfrac{\partial \xi_i^T}{\partial \varphi_{i+1}}\xi_i - \dfrac{\partial \xi_{i+1}^T}{\partial \varphi_{i+1}}\xi_{i+1}$ (7.8)

$$\dot\theta_i = \frac{\partial F}{\partial \theta_i} = -<\frac{\partial \xi_i^T}{\partial \theta_i}\xi>_y$$ (7.9)

Learning phase: $\dot\lambda_i = \dfrac{\partial F}{\partial \lambda_i} = -\left\langle \dfrac{\partial \xi_i^T}{\partial \lambda_i}\xi \right\rangle_p - (1 + \lambda_i)^{-1}$ (7.10)

The hierarchy of the model is shown in Fig. 7.1.

7.3 RESTRICTED BOLTZMANN MACHINES

The restricted Boltzmann machine (RBM)[6] is an important model in the field of deep learning, and it is also one of the basic units of many deep networks. In recent years, with the emergence of contrast dispersion fast learning algorithms, the research and application of RBM have been set off in the field of machine learning[7]. After learning effectively, RBM can provide the analytic representation of training data distribution, which is a generation model that can sample from the sample distribution.

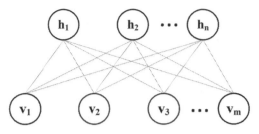

FIGURE 7.2   The restricted Boltzmann machine[7].

7.3.1 RBM model

RBM is a kind of neural network model with two-layer structure, symmetrical connection, and no self-feedback. Its structure is shown in Fig. 7.2. The first layer is composed of observation data variables (visible units), and the second layer is composed of hidden variables (hidden units). The visible layer and hidden layer are fully connected, but there is no connection between the visible layer and the hidden layer.

RBM can be defined in the form of energy function. Assuming that RBM contains m visible units and n hidden units, its energy function can be defined as:

$$E(v, h) = - \sum_{i=1}^{m} \sum_{j=1}^{n} w_{ij} v_i h_j - \sum_{i=1}^{m} b_i v_i - \sum_{j=1}^{n} c_j h_j \qquad (7.11)$$

where v_i, h_j are the states of visible unit i and hidden unit j, b_j and c_i are their offsets, w_{ij} is the connection weight between visible unit i and hidden unit j, and $\theta = \{W, b, c\}$ are RBM model parameters.

Based on this energy function, the joint concept distribution $p(v, h)$ is obtained by:

$$p(v, h) = \frac{1}{Z} e^{-E(v,h)} \qquad (7.12)$$

where Z is the partition function, which is the energy accumulation of all visible and implicit elements and can be defined by:

$$Z = \sum_{v,h} e^{-E(v,h)} \qquad (7.13)$$

Thus, the marginal probability distribution of the visible element v can be obtained by the following:

$$p(v) = \frac{1}{Z} \sum_h e^{-E(v,h)} \tag{7.14}$$

The training process of RBM aims at solving the value of model parameter θ, which can be obtained by maximizing the maximum likelihood function on the training set.

$$\theta^* = \arg\max_\theta L(\theta|v) = \arg\max_\theta \log p(v|\theta) \tag{7.15}$$

The gradient descent method is used to solve the above likelihood function:

$$\frac{\partial \log L(\theta|v)}{\partial \theta} = -\sum_h p(h|v)\frac{\partial E(v, h)}{\partial \theta} + \sum_{v,h} p(v, h)\frac{E(v, h)}{\partial \theta} \tag{7.16}$$

where $p(h|v)$ is the probability distribution of hidden layer when the visible element is known, which is easy to calculate. $p(v,h)$ represents the joint probability distribution of visible and hidden elements. Due to the existence of the normalization factor Z, it can not be directly calculated, and its approximate value can only be obtained by sampling.

7.3.2 Classification restricted Boltzmann machine

The classification restricted Boltzmann machine (ClassRBM)[8] is an improvement on the restricted Boltzmann machine. It establishes a three-layer RBM network of visible layer, hidden layer, and output layer for training data, so as to establish joint probability distribution between input unit x, hidden unit h, and class label y, and realize direct classification of RBM. The network structure diagram is shown in Fig. 7.3. If there are m nodes in visible layer, n nodes in hidden layer, and C output nodes in output layer, then each visible node is only related to n hidden layers, each hidden layer node is only related to m visible nodes, and C output nodes, each output node is only related to n hidden layer nodes, the connection weight matrix between visible layer and hidden layer is

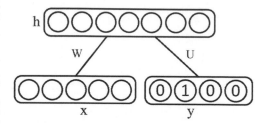

FIGURE 7.3 Illustration of the classification restricted Boltzmann machine[8].

W, and the connection between hidden layer and output layer is W. The weight matrix of is U.

The energy function of ClassRBM can be defined as:

$$E(y, x, h) = - \sum_{i=1}^{m} \sum_{j=1}^{n} x_i W_{ij} h_j - \sum_{i=1}^{m} b_i x_i - \sum_{j=1}^{n} c_j h_j - \sum_{k=1}^{C} d_k y_k$$
$$- \sum_{j=1}^{n} \sum_{k=1}^{C} h_j U_{jk} y_k \tag{7.17}$$

where m, n, C are the number of nodes in visible layer, hiden layer and output layer; b_i is the offset of the visible unit i, c_j is the offset of the hidden unit j, d_k is the offset of the output layer node k; W is the connection weight matrix between the visible layer and the hidden layer, and U is weight matrix between the hidden layer and output layer.

According to the energy function, the joint density can be defined as follows:

$$p(y, x, h) = \frac{\exp(-E(y, x, h))}{Z} \tag{7.18}$$

Similarly, it is impossible to directly calculate $p(y, x, h)$, but the approximate value can be obtained by using Gibbs sampling in ClassRBM, which can be alternately carried out by giving the hidden layer value of visible layer sampling or the value of visible layer sampling given hidden layer. These conditional distributions are very simple when a visual layer is given.

$$p(h|y, x) = \prod_j p(h_j|y, x) \tag{7.19}$$

where $p(h_j = 1|y, x) = sigm(c_j + U_{jy} + \Sigma_i W_{ji}x_i)$, $sigm(\alpha) = 1/(1 + exp(-\alpha))$ is the sigmoid activation function.

When a hidden layer is given, there are:

$$p(x|h) = \prod_i p(x_i|h), \text{ where } p(x_i = 1|h) = sigm(b_i + \sum_j W_{ji}h_j) \tag{7.20}$$

Thus, $p(y|h)$ can be calculated and classified:

$$p(y|h) = \frac{\exp(d_y + \Sigma_j U_{jy}h_j)}{\Sigma_{y^* \in \{1,2,...,C\}} \exp(d_{y^*} + \Sigma_j U_{jy^*}h_j)} \tag{7.21}$$

where $y*$ denotes all categories.

7.4 MEMORY MODEL BASED ON FREE ENERGY THEORY AND CLASSIFICATION CONSTRAINED BOLTZMANN MACHINE

The classification restricted Boltzmann machine ClassRBM can estimate the joint probability $p(y, x, h)$. This probability generation model has a similar form to the probability generation model proposed by Friston. Based on this, this chapter realizes the free energy of the system based on the ClassRBM model Minimize and establishes a Visual Memory Model based on Free Energy (FEVMM). In the ClassRBM network model, each node can store energy, and the correlation between variables determines the level of energy. When the system becomes stable, the sum of energy of all nodes is the minimum.

7.4.1 Definition of free energy function

The traditional restricted Boltzmann machine is a probabilistic neural network, which is used to learn the probability distribution between the visible layer and the hidden layer. In order to better describe the visual information of the real world, we use the visible unit which satisfies the Gaussian distribution. In this case, the energy function of the model is:

$$E(y, x, h) = \sum_{i=1}^{m} \frac{(x_i - b_i)^2}{2\sigma_i^2} - \sum_{i=1}^{m}\sum_{j=1}^{n} \frac{x_i}{v_i} W_{ij} h_j - \sum_{j=1}^{n} c_j h_j - \sum_{k=1}^{c} d_k y_k$$

$$- \sum_{j=1}^{n}\sum_{k=1}^{c} h_j U_{jk} y_k \tag{7.22}$$

where b_i is the offset of visible node i, σ_i is the standard deviation of Gaussian noise of visible node i, c_j is the offset of hidden layer node j, and d_k is the offset of output layer node k.

If the energy function is regarded as the objective function and the partial derivative of the objective function is obtained, the gradient descent method can be used to solve the problem. The energy of all nodes can be accumulated, and the accumulated result can be used to measure the energy of the whole network. But in this way, we must enumerate all the encoded samples (the values of hidden nodes and output nodes) corresponding to each sample, so as to calculate energy and consume a lot of computing resources. In order to solve this problem, the probability model is introduced, and then the energy is transformed into the expression of free energy through probability distribution to find the minimum free energy.

According to the energy function, the free energy function of the system is defined as:

$$F(y, x) = - \log \sum_{h} e^{-E(y,x,h)} \tag{7.23}$$

Thus, $p(y, x)$ can be rewritten as:

$$p(y, x) = \frac{e^{-F(y,x)}}{Z} \tag{7.24}$$

where Z is the partition function and can be defined as:

$$Z = \sum_{y,x,h} e^{-E(y,x,h)} \tag{7.25}$$

Then logarithm operation is taken on both sides of Equation (6.24) to obtain:

$$\log p(y, x) = -F(y, x) - \log Z \qquad (7.26)$$

It can be seen that the free energy can be measured by $\log p(y, x)$. The larger the $p(y, x)$ is, the smaller the free energy is:

$$\sum \log p(y, x) = -\sum F(y, x) - \sum \log Z \qquad (7.27)$$

The left term of Equation (7.27) is the logarithmic sum of probability $p(y, x)$, that is, the logarithmic likelihood function; the first term on the right is the negative value of the total free energy of the whole network, and the second term is a constant, which can be ignored. The relationship between the sum of free energy of a physical system and the logarithmic sum of $p(y, x)$ is obtained. When the sum of free energy of a system is the minimum, $\Pi p(y, x)$ would be the maximum. Therefore, the parameters obtained by maximum likelihood estimation can minimize the total free energy of RBM system, and the distribution of input data is also best fitted.

7.4.2 Model learning

When the free energy of the system reaches the minimum, the likelihood function gets the maximum value. Therefore, the maximum likelihood estimation can be used to train the model parameters, and the optimal parameters obtained can make the sum of free energy of the system minimum. The process of model training is to define an objective function and minimize the objective function for all samples in the training set $D_{train} = \{(x_t, y_t)\}$. The model parameters to be learned are $\Theta = \{b, c, d, W, U\}$.

The maximum likelihood function is defined by the joint probability distribution of x and y:

$$L_{gen}(D_{train}) = -\sum_{t=1}^{|D_{train}|} \log p(y_t, x_t) \qquad (7.28)$$

This is the most commonly used training objective function of RBM model, that is, the training method of generating model, and the likelihood function can be decomposed into the following forms:

$$L_{gen}(D_{train}) = - \sum_{t-1}^{\backslash D_{train}\backslash} (\log p(y_t|x_t) + \log p(x_t))$$

$$= - \sum_{t=1}^{|D_{train}|} \log p(y_t|x_t) - \sum_{t=1}^{|D_{train}|} \log p(x_t) \tag{7.29}$$

Since supervised learning mainly obtains a good prediction according to the given input, ignores the unsupervised part of the generating objective function, only focuses on the supervised part, and models the posterior conditional distribution. This training method is called discriminant training, and the objective function is defined as:

$$L_{disc}(D_{train}) = - \sum_{t=1}^{|D_{train}|} \log p(y_t|x_t) \tag{7.30}$$

This training objective is similar to the objective function used in feed-forward neural network, and the output of the network can be regarded as an estimation of $p(y|x)$.

The related research shows that the model obtained by generative training is more standardized than that by discriminant training, and adding generative training target to discriminant training target can be used as a normalization method for discriminating training target. Therefore, a hybrid training method can be adopted, that is, both generative model and discriminant model are used for model training. The contribution of the two models to the training process can be determined by the weight $\alpha(0 < \alpha < 1)$, so the following can be gotten:

$$L_{hybrid}(D_{train}) = \alpha L_{gen}(D_{train}) + (1 - \alpha)L_{disc}(D_{train})$$

$$= - \alpha \sum_{t=1}^{|D_{train}|} \log p(x_t) - \sum_{t=1}^{|D_{train}|} \log p(y_t|x_t) \tag{7.31}$$

The Friston free energy formula described in Equation (6.31) is similar to that in Equation (6.4). The first term indicates that the generation model can simulate the input distribution better by minimizing the objective function, and the second term represents that the posterior probability can be well estimated by minimizing the objective function.

When the above objective function reaches the minimum, the system reaches the stable state, that is, the free energy of the system reaches the minimum.

7.4.2.1 Generative model training

The gradient descent method is used to solve the training objective function defined in Equation (6.28) to calculate the gradient of *log p(y_t, x_t)* for arbitrary model parameter θ:

$$
\begin{aligned}
\frac{\partial \log p(y,x)}{\partial \theta} &= -\frac{\sum_h e^{-E(y,x,h)} \frac{\partial E(y,x,h)}{\partial \theta}}{\sum_h e^{-E(y,x,h)}} + \frac{\sum_{x,y,h} e^{-E(x,y,h)} \frac{\partial E(y,x,h)}{\partial \theta}}{\sum_{x,y,h} e^{-E(y,x,h)}} \\
&= -\sum_h p(h|x,y) \frac{-\partial E(y,x,h)}{\partial \theta} + \sum_{x,y,h} p(y,x,h) \frac{\partial E(y,x,h)}{\partial \theta} \\
&= -\mathbb{E}_{p(h|x,y)} \left[\frac{\partial E(y,x,h)}{\partial \theta} \right] + \mathbb{E}_{p(y,x,h)} \left[\frac{\partial E(y,x,h)}{\partial \theta} \right]
\end{aligned}
$$

(7.32)

The first term of the last equation in Equation (6.32) is the expectation of the function $\frac{\partial E(y,x,h)}{\partial \theta}$ under the probability $p(h|x, y)$, and the second term is the expectation of the function $\frac{\partial E(y,x,h)}{\partial \theta}$ under the probability $p(y, x, h)$. The first term is equal to the expected value of the energy function of the input sample. The second term is the expected value of the energy function of the sample data generated by the model. The joint distribution $p(y, x, h)$ is difficult to obtain, but the joint distribution can be fitted by Gibbs sampling according to the conditional distribution, and the conditional distribution is easier to obtain.

Given the visible layer and output layer, there are:

$$
p(h|x, y) = \prod_j p(h_j|x, y), \; p(h_j|x, y) = sigm\left(\sum_i W_{ij} x_i + \sum_k U_{kj} y_k + c_j \right)
$$

(7.33)

where $sigm(\alpha) = \frac{1}{1 + \exp(-\alpha)}$ is a logisticfunction.

Given the hidden layer, there are:

$$p(y|h) = \frac{\exp(d_y + \sum_j U_{jy} h_j)}{\sum_{y^*} \exp(d_{y^*} + \sum_j U_{jy^*} h_j)}, \; p(y_k|h) = sigm\left(\sum_j U_{kj} h_j + d_k\right)$$

(7.34)

$$p(x|h) = \prod_i p(x_i, h), \; p(x_i, h) = N\left(x_i \sum_j W_{ij} h_j + b_i, \sigma_i^2\right)$$ (7.35)

where $N(x|\mu, \sigma^2)$ is a Gaussian distribution density function with mean value μ and variance σ^2.

In order to speed up the sampling process, CD-k algorithm is used to fit the joint probability distribution $p(y, x, h)$. Firstly, the training sample is taken as the initial value, and then k-step Gibbs sampling is performed. Generally, k is set to 1. The learning process is shown in Algorithm 7.1.

Algorithm 7.1   LEARNING ALGORITHM FOR GENERATING MODEL PARAMETERS

Input: the t-th training sample (x_t, y_t), learning rate λ
Output: parameters $\Theta = \{b, c, d, W, U\}$
CD-k sampling (for each training sample):
Negative phase:

$$y^0 \leftarrow y_t, \; x^0 \leftarrow x_t, \; \hat{h}^0 \leftarrow sigm(c + Wx^0 + Uy^0)$$

Positive stage:

$$h^0 \sim p(h|y^0, x^0), \; y^1 \sim p(y|h^0), \; x^1 \sim p(x|h^0)$$

$$\hat{h}^1 \leftarrow sigm(c + Wx^1 + Uy^1)$$

Parameter update: for each parameter θ

$$\theta^t = \theta^{t-1} + \lambda\left(\frac{\partial}{\partial \theta}E(y^1, x^1, \hat{h}^1) - \frac{\partial}{\partial \theta}E(y^0, x^0, \hat{h}^0)\right)$$

The probability distribution in accordance with $p(y, x, h)$ is sampled through the above process, and the network parameters are gradually optimized to obtain the optimal solution of the parameters.

7.4.2.2 Training of discriminative model

Using the discriminative training objective function defined in Equation (7.30) by gradient descent method, the gradient of $log\ p(y_t,\ x_t)$ for arbitrary model parameter θ is calculated by:

$$\frac{\partial \log p\,(y_t|x_t)}{\partial \theta} = -\mathbb{E}_{h|y_t,x_t}\left[\frac{\partial}{\partial \theta}E\,(y_t,\ x_t,\ h)\right] + \mathbb{E}_{y,h|x}\left[\frac{\partial}{\partial \theta}E\,(y,\ x,\ h)\right]$$

$$(7.36)$$

Specifically, for the parameter d:

$$\frac{\partial \log p\,(y_t|x_t)}{\partial d_y} = 1_{y=y_t} - p\,(y|x_t), \quad \forall\ y \in \{1,\ \cdots,\ C\} \qquad (7.37)$$

For other parameters $\theta \in \{c,\ U,\ W\}$:

$$\frac{\partial log\ p\,(y_t|x_t)}{\partial \theta} = \sum_j sigm\,(o_{y,j}\,(x_t))\frac{\partial o_{y,j}\,(x_t)}{\partial \theta} - \sum_{j,y*} sigm\,(o_{y*j}\,(x_t))p\,(y*|x_t)$$

$$\frac{\partial o_{y*j}\,(x_t)}{\partial \theta} \qquad\qquad\qquad\qquad (7.38)$$

Because the input bias is not involved in the calculation of $p\,(y|x)$, the gradient of parameter b is 0.

7.4.3 Recall

Given the test sample x, the posterior probability of Class y is:

$$p\,(y|x) = \frac{exp\,(d_y + \sum_j softplus\,(c_j + U_{jy} + \sum_i W_{ij}x_i))}{\sum_{y*} exp\,(d_{y*} + \sum_j softplus\,(c_j + U_{jy*} + \sum_i W_{ij}x_i))} \qquad (7.39)$$

Then, the posterior probability $p(y|\ x)$ is calculated for each category, and the class i^* with the maximum posterior probability is calculated by:

$$i* = arg\ max\,\{p_i\} \qquad (7.40)$$

where p_i represents the posterior probability of the ith category.

If the posterior probability of the i^* class exceeds the pre-defined threshold Φ, the sample is considered to have learned and classified into the i^* class; otherwise, the sample x is considered to have not been learned and is a new sample.

$$\text{Re } sult = \begin{cases} new & p_{i*} < \Phi \\ i * class & p_{i*} \geq \Phi \end{cases} \tag{7.41}$$

7.5 EXPERIMENTAL RESULTS

7.5.1 Experimental setup

In order to verify the effectiveness of the method proposed in this chapter, experiments are carried out on Corel-1000 and UIUC sports standard databases and compared with SVM[9], SOINN[10], and IPAMM[11]. For each database, training samples and test samples were randomly selected, all images kept the original size, and different numbers of images (5, 10, 20, 30, 40) were selected for training and the remaining images were used for testing. First, the image features are extracted by using the deep convolution network, and a 4096-dimensional feature vector is obtained from each image, and then the extracted features are input into FEVMM, SVM, SOINN, and IPAMM models for classification and recognition. In the FEVMM model proposed in this chapter, the number of hidden units is 100, the iterations are 50 times, the learning rate is 0.05, the weight of hybrid model is $\alpha = 0.01$, and the threshold Φ is set to 0.5.

7.5.2 Corel-1000 dataset

Corel-1000 image dataset contains 10 kinds of objects (Africa, beaches, buildings, buses, dinosaurs, elephants, flowers, foods, horses, mountains), and each class contains 100 images. Some example images are shown in Fig. 7.4.

7.5.2.1 Changes of free energy during training

In order to observe the change of system free energy during the process of model learning, 10 and 20 images are selected from each category as training set, and the remaining images are used as verification set for

FIGURE 7.4 Sample images from Corel dataset.

model training. The number of implicit elements is 100 and the iteration is 200 times. During the training process, the mean square error (MSE) and free energy of the model with the number of iterations are shown in Fig. 7.5 and Fig. 7.6, respectively.

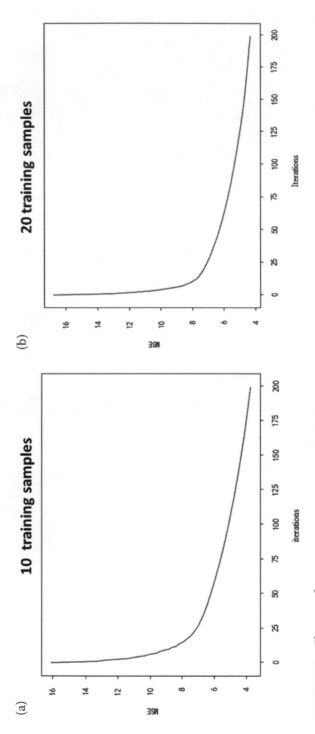

FIGURE 7.5   Change of mean square error.

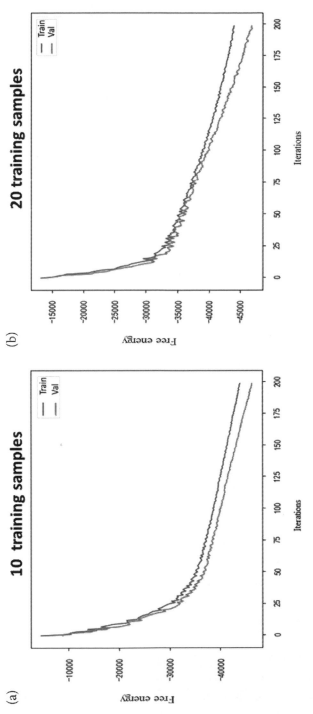

FIGURE 7.6 Change of free energy of the system.

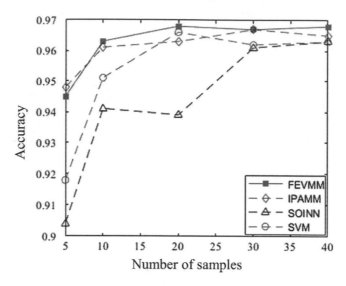

FIGURE 7.7   Performance comparison on Corel dataset.

During the process of model training, with the increase of the number of iterations, the mean square error MSE decreases gradually, and the free energy of the system also decreases gradually. After a period of training, the MSE and free energy of the system tend to be stable, and the free energy of the system is the minimum. It can be seen that the process of model training is the process of minimizing prediction error and free energy, which is consistent with Friston's free energy theory.

7.5.2.2 Classification performance

Fig. 7.7 shows the classification performance comparison results of the FEVMM model and SVM, SOINN, and IPAMM models under different number of training samples. The experimental results show that the FEVM model is obviously better than SOINN and SVM models. Compared with IPAMM, the proposed method is slightly worse than the IPAMM model when the number of samples is less than 5. When the number of training samples is more than 10, the FEVMM model has better classification performance.

7.5.3 UIUC Sports dataset

UIUC Sports dataset consists of eight sports categories including badminton, bocce, croquet, polo, rockclimbing, rowing, sailing, and

FIGURE 7.8 Sample images from UIUC Sports dataset. From top to bottom, categories are respectively badminton, bocce, croquet, polo, rockclimbing, rowing, sailing, and snowboarding.

snowboarding, as shown in Fig. 7.8. Each category contains a variety number of pictures, ranging from 137 to 250.

The classification and comparison results of several methods are shown in Fig. 7.9. On the whole, the FEVMM model is better than SOINN and SVM model, and better than the IPAMM model when there are more samples. It can be seen that this method can achieve

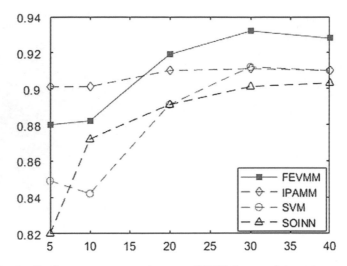

FIGURE 7.9   Performance comparison on UIUC Sports dataset.

better classification performance when there are more training samples, while the IPAMM model has better performance in the case of less training samples.

7.6 CONCLUSIONS

Based on the understanding of Friston's free energy theory about brain perception and reasoning, this chapter combines free energy theory with the restricted Boltzmann machine to establish a visual information memory model and realize the storage and extraction of visual information. Experimental results show that the classification performance of the proposed method is better than that of SOINN and SVM under a different number of training samples, and the classification effect is better than the IPAMM model when there are more training samples (more than 10).

REFERENCES

1. Friston, K. 2005. A theory of cortical responses. *Philosophical Transactions of the Royal Society of London B: Biological Sciences* 360(1456):815–836.
2. Friston, K. J., Stephan, K. E. 2007. Free-energy and the brain. *Synthese* 159 (3):417–458.
3. Bogacz, R. 2017. A tutorial on the free-energy framework for modelling perception and learning. *Journal of mathematical psychology* 76: 198–211.

4. Friston, K., Kilner, J., Harrison, L. 2006. A free energy principle for the brain. *Journal of Physiology-Paris* 100(1–3): 70–87.

5. Friston, K. 2005. A theory of cortical responses. *Philosophical Transactions of the Royal Society of London B: Biological Sciences* 360(1456):815–836.

6. Salakhutdinov, R., Hinton, G. Deep Boltzmann machines. Proceedings of the International Conference on Artificial Intelligence and Statistics: 448–455.

7. Zhang, C., Ji, N., Wang, G. 2015. Restricted Boltzmann machine. *Journal of Engineering Mathematics* 32(2): 159–173.

8. Larochelle, H., Mandel, M., Pascanu, R., et al. 2012. Learning algorithms for the classification restricted Boltzmann machine. *Journal of Machine Learning Research* 13(Mar):643–669.

9. Chang, C. C., Lin, C. J. LIBSVM: a library for support vector machines. *ACM transactions on intelligent systems and technology (TIST)* 2(3):27.

10. Furao, S., Hasegawa, O. 2006. An incremental network for on-line unsupervised classification and topology learning. *Neural Netw* 19(1): 90–106.

11. Deng, L., Gao, M., Wang, Y. 2018. Increment learning and rapid retrieval of visual information based on pattern association memory. *Neural Processing Letters* 48(3):1597–1610.

APPENDIX

```
#Programmed by python 2.7
```

1. example.py

```python
# encoding:utf-8
from pylab import mpl
mpl.rcParams['font.sans-serif'] = ['SimHei']
mpl.rcParams['axes.unicode_minus'] = False
import morb
#from morb import rbms, stats, updaters, trainers,
monitors, units, parameters
from morb import rbms, stats, updaters, trainers,
monitors, units, parameters, prediction, objectives
import theano
import theano.tensor as T
import sys
import numpy as np
```

```
import gzip, cPickle
import scipy.io as sio

import matplotlib.pyplot as plt
plt.ion()
from sklearn import preprocessing
from utils import generate_data, get_context,
one_hot
from confmat import Confmat
import json, sys, os, time, os.path, math
numpy_rng = np.random.RandomState(23355)

theano.config.floatX = 'float32'
mode = None
reload(sys)
sys.setdefaultencoding('utf8')
# load data
print ">> Loading dataset..."


#load .mat file
train_data = []
train_labels = []
m = sio.loadmat("datasets/Corel_20.mat")
train_set_x=m['XTrain']
#train_set_x=zip(*train_set_x)
train_set_x=np.array(train_set_x).T

train_set_y =m['TrainLabel']
s=len(train_set_y)
train_set_y.shape=(1,s)
train_set_y.transpose()
train_set_y=[x-1 for x in train_set_y]


valid_set_x=train_set_x
valid_set_y =train_set_y
```

```
test_set_x=m['XTest']
test_set_x=np.array(test_set_x).T

test_set_y =m['TestLabel']
s=len(test_set_y)
test_set_y.shape=(1,s)
test_set_y.transpose()
test_set_y=[x-1 for x in test_set_y]

# normalize the data attributes
'''
scaler_train  =  preprocessing.StandardScaler().fit
(train_set_x)
train_set_x = scaler_train.transform(train_set_x)

scaler_test   =  preprocessing.StandardScaler().fit
(test_set_x)
test_set_x= scaler_test.transform(test_set_x)
'''
# Z-score normalise

global_mu = np.mean(train_set_x)
global_sigma = np.std(train_set_x)
train_set_x -= global_mu
train_set_x /= (0.25* global_sigma)   #0.25

#global_mu1 = np.mean(test_set_x)
#global_sigma1 = np.std(test_set_x)
test_set_x -= global_mu
test_set_x /= (0.25 * global_sigma)


# convert labels to one hot representation
train_set_y_oh = one_hot(np.atleast_2d(train_
set_y).T)
valid_set_y_oh = one_hot(np.atleast_2d(valid_
set_y).T)
test_set_y_oh = one_hot(np.atleast_2d(test_set_y).T)
```

```
# dim 0 = minibatches, dim 1 = units, dim 2 = states


train_set_y_oh = train_set_y_oh.reshape((train_-
set_y_oh.shape[0], train_set_y_oh.shape[1]))
valid_set_y_oh = valid_set_y_oh.reshape((valid_
set_y_oh.shape[0], valid_set_y_oh.shape[1]))
test_set_y_oh = test_set_y_oh.reshape((test_set_
y_oh.shape[0], test_set_y_oh.shape[1]))


train_set_x = train_set_x[:10000]
train_set_y_oh = train_set_y_oh[:10000]
valid_set_x = valid_set_x[:1000]
valid_set_y_oh = valid_set_y_oh[:1000]

# make the sets a bit smaller for testing purposes


#parameters
visible_maps = train_set_x.shape[1]
hidden_maps =200
n_states = train_set_y_oh.shape[1]
epochs=300
learning_rate=0.05
beta=0.01
context_units = []
mb_size=40


train_data = train_set_x
train_labels = train_set_y
test_data = test_set_x
test_labels = test_set_y


###############################################
# CONSTRUCT RBM
```

```
###############################################
print ">> Constructing RBM..."
weight_w_std=4*np.sqrt(6./(hidden_maps
+visible_maps))
weight_u_std = 4*np.sqrt(6./(hidden_maps+n_states))
initial_W = np.asarray( numpy_rng.uniform(
    low=-weight_w_std,
    high=weight_w_std,
    size=(visible_maps,hidden_maps)),
    dtype=theano.config.floatX)
initial_U = np.asarray(numpy_rng.uniform(
    low=-weight_u_std,
    high=weight_u_std,
    size=(n_states, hidden_maps)),
    dtype=theano.config.floatX)

initial_bv = np.zeros(visible_maps, dtype = theano.
config.floatX)
initial_by = np.zeros(n_states, dtype = theano.
config.floatX)
initial_bh = np.zeros(hidden_maps, dtype = theano.
config.floatX)
rbm = morb.base.RBM()
#rbm.v = units.BinaryUnits(rbm, name='v') # visibles
rbm.v = units.GaussianUnits(rbm, name='v') # visibles
rbm.h = units.BinaryUnits(rbm, name='h') # hiddens
rbm.y = units.SoftmaxUnits(rbm, name='y')

pmap = {
 "W": theano.shared(value=initial_W, name="W"),
 "bv": theano.shared(value=initial_bv, name="bv"),
 "bh": theano.shared(value=initial_bh, name="bh"),
 "U" : theano.shared(value=initial_U, name="U"),
 "by": theano.shared(value=initial_by, name="by")
}

rbm.W = parameters.ProdParameters(rbm, [rbm.v,
rbm.h], 'W', name='W')
```

```
rbm.bv = parameters.BiasParameters(rbm, rbm.v,
'bv', name='bv')
rbm.bh = parameters.BiasParameters(rbm, rbm.h, 'bh',
name='bh')
rbm.U = parameters.ProdParameters(rbm, [rbm.y,
rbm.h], 'U', name='U')
rbm.by = parameters.BiasParameters(rbm, rbm.y, 'by',
name='by')

initial_vmap = { rbm.v: T.matrix('v'), rbm.y:
T.matrix('y')}
print rbm

dlo = objectives.discriminative_learning_objective
(rbm, \
    visible_units = [rbm.v], \
    hidden_units = [rbm.h], \
    label_units = [rbm.y], \
    vmap = initial_vmap,
    pmap = pmap)


# try to calculate weight updates using CD-1 stats
k_train = 1
k_eval = 1
print ">> Constructing contrastive divergence
updaters..."
s = stats.cd_stats(rbm, initial_vmap, pmap, visible_
units=[rbm.v, rbm.y], hidden_units=[rbm.h], context_
units=context_units, k=k_train, mean_field_for_stats=
[rbm.v, rbm.y], mean_field_for_gibbs=[rbm.v, rbm.y])
persistent_vmap= dict((h, h.activation(initial_
vmap, pmap)) for h in [rbm.h])
pmap_regular_plus_fast=pmap;
#s=stats.fast_pcd_stats(rbm, initial_vmap,
persistent_vmap, pmap, pmap_regular_plus_fast,
visible_units=[rbm.v, rbm.y], hidden_units=[rbm.h],
```

```
context_units=context_units,mean_field_for_stats=
[rbm.v, rbm.y], mean_field_for_gibbs=[rbm.v, rbm.y])
print "Stats eval"
s_eval = stats.cd_stats(rbm, initial_vmap, pmap,
visible_units=[rbm.v, rbm.y], hidden_units=[rbm.h],
context_units=context_units, k=k_eval) # , mean_
field_for_gibbs=[rbm.v], mean_field_for_stats=[rbm.v,
rbm.h])

umap = {}
learning_rate_with_decay = theano.shared(value=np.
cast[theano.config.floatX](learning_rate),
name='learning_rate')
decayed_beta = theano.shared(value=np.cast['float32']
(beta), name='beta')

sparsity_cost = 0.001
sparsity_targets = { rbm.h: 0.1}
for var in rbm.variables:
 pu = pmap[var]
 '''
 if var=='W' or var=='bh' :
   pu += learning_rate_with_decay * ( decayed_beta *
updaters.CDUpdater(rbm, var, s) + (1 - decayed_beta)
* updaters.GradientUpdater(dlo, var, pmap=pmap))\
       +learning_rate_with_decay * sparsity_cost *
updaters.SparsityUpdater(rbm,    var,    sparsity_-
targets, s)
   print "sparsity"
 else:
   pu += learning_rate_with_decay * ( decayed_beta *
updaters.CDUpdater(rbm, var, s) + (1 - decayed_beta)
* updaters.GradientUpdater(dlo, var, pmap=pmap))
 '''
 #pu += learning_rate_with_decay * (decayed_beta *
updaters.CDUpdater(rbm, var, s) + (1 - decayed_beta)
* updaters.GradientUpdater(dlo, var, pmap=pmap))
 pu += learning_rate_with_decay * (
```

```
 decayed_beta * updaters.CDUpdater(rbm, var, s) +
updaters.GradientUpdater(dlo, var, pmap=pmap))
 umap[pmap[var]] = pu
print ">> Compiling functions..."
t = trainers.MinibatchTrainer(rbm, umap)
mse_v =monitors.reconstruction_mse(s, rbm.v)
mse_y = monitors.reconstruction_mse(s, rbm.y)
m_data = s['data'][rbm.v]
m_model = s['model'][rbm.v]
#e_data = rbm.energy(s['data'], pmap).mean()
#e_model = rbm.energy(s['model'], pmap).mean()
#free_energy = T.mean(rbm.free_energy([rbm.h],
s['data'])) # take the mean over the minibatch.

e_data = rbm.free_energy([rbm.h],s['data'], pmap).
mean()
e_model = rbm.free_energy([rbm.h],s['model'], pmap).
mean()
h_data = s['data'][rbm.h]
h_model = s['model'][rbm.h]

v_data = s['data'][rbm.v]
v_model = s['model'][rbm.v]

train = t.compile_function(initial_vmap, mb_size=
mb_size, monitors=[mse_v, mse_y, e_data, e_model,
h_data, h_model, v_model, dlo, v_data], name=
'train', mode=mode)
evalt = t.compile_function(initial_vmap, mb_size=
mb_size, monitors=[mse_v, mse_y, e_data, e_model,
h_data, h_model, v_model, dlo, v_data], name='eval',
train=False, mode=mode)

print ">> Compiling prediction..."

predict = prediction.label_prediction(rbm, initial_
vmap, pmap, \
  visible_units = [rbm.v], \
```

```
        label_units = [rbm.y] , \
        hidden_units = [rbm.h] ,
        context_units = context_units,
        mb_size=mb_size, mode=mode,
        only_activation = True)

    print ">> Compiling evaluation..."
    umap_e = {}
    for var in rbm.variables:
        umap_e[pmap[var]] = pmap[var] + learning_rate *
    updaters.CDUpdater(rbm, var, s_eval)
    t_eval = trainers.MinibatchTrainer(rbm, umap_e)
    msev_eval = monitors.reconstruction_mse(s_eval,
    rbm.v)
    msey_eval = monitors.reconstruction_mse(s_eval,
    rbm.y)
    evaluate = t_eval.compile_function(initial_vmap,
    mb_size=mb_size, monitors=[msev_eval, msey_eval,
    s_eval['data'][rbm.v], s_eval['model'][rbm.v], s_
    eval['data'][rbm.h], s_eval['model'][rbm.h]], name
    ='evaluate', train=False, mode=mode)


    ###############################################
    # TRAINING
    ###############################################

    print ">> Training for %d epochs..." % epochs
    print

    stats = {}

    def record_stat(name, epoch, value):
        stats.setdefault(name, []).append((epoch, float
        (value)))

    def confmat(x_set, y_set, x_set_border = None):
        expected = np.argmax(y_set, axis=1)
```

```
    predicted = np.concatenate([np.argmax(y, axis=1)
    for y, in predict({ rbm.v: x_set })])
    return Confmat(expected, predicted)
def record_stat(name, epoch, value):
    stats.setdefault(name, []).append((epoch, float
    (value)))

## TRAIN
start_time = time.time()
test_result=[]

mses_train_so_far = []
mses_train_so_far1 = []
mses_valid_so_far = []
edata_train_so_far = []
emodel_train_so_far = []
edata_so_far = []
emodel_so_far = []

for epoch in range(epochs):
    print "epoch: %6d, time = %.2f s" % (epoch, time.
    time() - start_time)
    var_prev_vals = {}
for var in rbm.variables:
    var_prev_vals[var] = pmap[var].get_value()

 costs = [(mv, my, ed, em, hd, hm, vm, d, vd) for mv, my,
ed, em, hd, hm, vm, d, vd in train({rbm.v: train_set_x,
rbm.y: train_set_y_oh}, shuffle_batches_rng=numpy_
rng)]

var_diffs = {}
 for var in rbm.variables:
   var_diffs[var] = np.sum(np.abs(var_prev_
   vals[var] - pmap[var].get_value()))
  record_stat("update_" + str(var), epoch, var_diffs
  [var])
print "epoch: %6d, updates since previous epoch: %s"
```

```
% (epoch, repr(var_diffs))


# average over minibatches
  msev_train = np.mean([mv for mv, my, ed, em, hd, hm,
  vm, d, vd in costs])
  msey_train = np.mean([my for my, my, ed, em, hd, hm,
  vm, d, vd in costs])
  edata_train = np.mean([ed for mv, my, ed, em, hd, hm,
  vm, d, vd in costs])
  emodel_train = np.mean([em for m, my, ed, em, hd, hm,
  vm, d, vd in costs])
  dlo_train = np.mean([d for m, my, ed, em, hd, hm, vm,
  d, vd in costs])

 print "epoch: %6d, train:  MSE v = %.6f, MSE y = %.6f,
data energy = %.2f, model energy = %.2f, dlo = %.2f" % (
  epoch, msev_train, msey_train, edata_train,
  emodel_train, dlo_train)

# record stats
  record_stat("msev_train", epoch, msev_train)
  record_stat("msey_train", epoch, msey_train)
  record_stat("edata_train", epoch, edata_train)
  record_stat("emodel_train", epoch, emodel_train)
  record_stat("dlo_train", epoch, dlo_train)
 if math.isnan(msev_train) or math.isinf(msev_train):
  print "NaN or inf"
  sys.exit()

## EVALUATE?
  eval_res = [(mv, my, ed, em, hd, hm, vm, d, vd) for mv,
  my, ed, em, hd, hm, vm, d, vd in  evalt({rbm.v:
  test_set_x, rbm.y: test_set_y_oh})]

# record stats
  msev_test = np.mean([mv for mv, my, ed, em, hd, hm,
  vm, d, vd in eval_res])
```

```python
    msey_test = np.mean([my for mv, my, ed, em, hd, hm,
    vm, d, vd in eval_res])
    edata_test = np.mean([ed for mv, my, ed, em, hd, hm,
    vm, d, vd in eval_res])
    emodel_test = np.mean([em for mv, my, ed, em, hd, hm,
    vm, d, vd in eval_res])
    dlo_test = np.mean([d for m, my, ed, em, hd, hm, vm,
    d, vd in eval_res])

record_stat("msev_test", epoch, msev_test)
record_stat("msey_test", epoch, msey_test)
record_stat("edata_test", epoch, edata_test)
record_stat("emodel_test", epoch, emodel_test)
record_stat("dlo_test", epoch, dlo_test)

# plotting

mses_train_so_far1.append(msey_test)
mses_train_so_far.append(msev_train)
#mses_train_so_far.append(msev_train)
mses_valid_so_far.append(msev_test)
edata_so_far.append(edata_test)
emodel_so_far.append(emodel_test)
edata_train_so_far.append(edata_train)
emodel_train_so_far.append(emodel_train)

print "epoch: %6d, test epoch, msev_test, msey_tes:
MSE v = %.6f, MSE y = %.6f, data energy = %.2f, model
energy = %.2f, dlo = %.2f" % (epoch, msev_test,
msey_test, edata_test, emodel_test, dlo_test)

## CLASSIFICATION
confmat_train = confmat(train_set_x, train_set_y_oh)
confmat_test = confmat(test_set_x, test_set_y_oh)

accuracy_train = confmat_train.accuracy()
accuracy_test = confmat_test.accuracy()
```

```
record_stat("accuracy_train", epoch, accuracy_train)
record_stat("accuracy_test", epoch, accuracy_test)
test_result.append(accuracy_test)
print "epoch: %6d, train accuracy = %.4f, test accuracy
= %.4f" % (epoch, accuracy_train, accuracy_test)
  print "epoch: %6d, confusion matrix train:" % (epoch)
#print confmat_train
  print "epoch: %6d, confusion matrix test:" % (epoch)
#print confmat_test
    print "train_result = %.4f, test_result = %.4f"
    %(accuracy_train,max(test_result))


plt.figure(1)
plt.plot(mses_train_so_far)
#plt.plot(mses_train_so_far, label=u'Train')
#plt.plot(mses_train_so_far1,label=u'Val')
#plt.title("MSE")
plt.xlabel(u"Interations")
plt.ylabel(u"MSE")
#plt.legend()
plt.draw()


plt.figure(2)
#plt.clf()
#plt.plot(edata_so_far, label=u'Val/ data')
#plt.plot(edata_train_so_far, label=u'Test/ data')
plt.plot(emodel_train_so_far)
#plt.plot(emodel_train_so_far, label=u'Train')
#plt.plot(emodel_so_far, label=u'Val')


#plt.title("Free energy")
plt.xlabel(u"Iterations")
plt.ylabel(u"Free energy")
plt.legend(loc=0)
plt.draw()


plt.pause(300)
```

2. rbms.py

```python
from morb.base import RBM
from morb import units, parameters

import theano
import theano.tensor as T
import numpy as np


### RBMS ###

class BinaryBinaryRBM(RBM): # the basic RBM, with
binary visibles and binary hiddens
    def __init__(self, n_visible, n_hidden):
        super(BinaryBinaryRBM, self).__init__()
        # data shape
        self.n_visible = n_visible
        self.n_hidden = n_hidden
        # units
        self.v = units.BinaryUnits(self, name='v') #
        visibles
        self.h = units.BinaryUnits(self, name='h') #
        hiddens
        # parameters
        self.pmap = {
            'W': theano.shared(value = self._initial_
            W(), name='W'),
            'bv': theano.shared(value = self._initial_
            bv(), name='bv'),
            'bh': theano.shared(value = self._initial_
            bh(), name='bh')
        }
        self.W = parameters.ProdParameters(self,
        [self.v, self.h], 'W', name='W') # weights
        self.bv = parameters.BiasParameters(self,
        self.v, 'bv', name='bv') # visible bias
```

```
        self.bh = parameters.BiasParameters(self,
        self.h, 'bh', name='bh') # hidden bias

    def _initial_W(self):
        return np.asarray( np.random.uniform(
            low  = -4*np.sqrt(6./(self.n_hidden+self.
            n_visible)),
            high = 4*np.sqrt(6./(self.n_hidden+self.
            n_visible)),
            size = (self.n_visible, self.n_hidden)),
            dtype = theano.config.floatX)

    def _initial_bv(self):
        return np.zeros(self.n_visible, dtype = theano.
        config.floatX)

    def _initial_bh(self):
        return np.zeros(self.n_hidden, dtype = theano.
        config.floatX)


class BinaryBinaryCRBM(BinaryBinaryRBM):
    def __init__(self, n_visible, n_hidden, n_
    context):
        super(BinaryBinaryCRBM, self).__init__
        (n_visible, n_hidden)
        # data shape
        self.n_context = n_context
        # units
        self.y = units.Units(self, name='y')
        #self.y = units.SoftmaxUnits(self, name='y')
        # context
        # parameters
        self.pmap['U'] = theano.shared(value = self._
        initial_U(), name='U')
        self.pmap['by'] = theano.shared(value = self._
        initial_by(), name='by')
        self.U = parameters.ProdParameters(self, [self.
```

```
    y, self.h], 'U', name='U') # context-to-hidden
    weights
  # self.U=parameters.AdvancedProdParameters(self,
  [self.y, self.h], [2, 1],'U', name='U') # context-
  to-hidden weights
   self.by =parameters.BiasParameters(self, self.y,
   'by', name='by') # context bias

  def _initial_U(self):
    # return np.zeros((self.n_context, self.n_
    hidden), dtype = theano.config.floatX)
    weight_u_std = 4 * np.sqrt(6. / (self.n_hidden +
    self.n_context))
    return np.asarray( np.random.uniform(
        low=-weight_u_std,
        high=weight_u_std,
        size=(self.n_context, self.n_hidden)),
        dtype=theano.config.floatX)

  def _initial_by(self):
    return np.zeros(self.n_context, dtype=theano.
    config.floatX)


class GaussianBinaryRBM(RBM): # Gaussian visible
units
  def __init__(self, n_visible, n_hidden):
    super(GaussianBinaryRBM, self).__init__()
    # data shape
    self.n_visible = n_visible
    self.n_hidden = n_hidden
    # units
    self.v = units.GaussianUnits(self, name='v')
    # visibles
    self.h = units.BinaryUnits(self, name='h') #
    hiddens
    # parameters
```

```
      parameters.FixedBiasParameters(self, self.v.
      precision_units)
      self.W = parameters.ProdParameters(self,
      [self.v, self.h], theano.shared(value = self._
      initial_W(), name='W'), name='W') # weights
      self.bv = parameters.QuadraticBiasParameters
      (self, self.v, theano.shared(value = self._
      initial_bv(), name='bv'), name='bv') # visible
      bias
      self.bh = parameters.BiasParameters(self,
      self.h, theano.shared(value = self._
      initial_bh(), name='bh'), name='bh')
      # hidden bias

  def _initial_W(self):
    return np.asarray( np.random.uniform(
        low  = -4*np.sqrt(6./(self.n_hidden+self.
        n_visible)),
        high = 4*np.sqrt(6./(self.n_hidden+self.
        n_visible)),
        size = (self.n_visible, self.n_hidden)),
        dtype = theano.config.floatX)

  def _initial_bv(self):
    return np.zeros(self.n_visible, dtype = theano.
    config.floatX)

  def _initial_bh(self):
    return np.zeros(self.n_hidden, dtype = theano.
    config.floatX)


class LearntPrecisionGaussianBinaryRBM(RBM):
  """
  Important: Wp and bvp should be constrained to be
  negative.
  """
  def __init__(self, n_visible, n_hidden):
```

```
    super(LearntPrecisionGaussianBinaryRBM,
    self).__init__()
    # data shape
    self.n_visible = n_visible
    self.n_hidden = n_hidden
    # units
    self.v = units.LearntPrecisionGaussianUnits
    (self, name='v') # visibles
    self.h = units.BinaryUnits(self, name='h')
    # hiddens
    # parameters
    self.Wm = parameters.ProdParameters(self,
    [self.v, self.h], theano.shared(value = self._
    initial_W(), name='Wm'), name='Wm') # weights
    self.Wp = parameters.ProdParameters(self,
    [self.v.precision_units, self.h], theano.
    shared(value = -np.abs(self._initial_W())/1000,
    name='Wp'), name='Wp') # weights
    self.bvm = parameters.BiasParameters(self, self.v,
    theano.shared(value = self._initial_bias(self.n_
    visible), name='bvm'), name='bvm') # visible bias
    self.bvp = parameters.BiasParameters(self,
    self.v.precision_units, theano.shared(value =
    self._initial_bias(self.n_visible),
    name='bvp'), name='bvp') # precision bias
    self.bh = parameters.BiasParameters(self, self.h,
    theano.shared(value = self._initial_bias(self.n_
    hidden), name='bh'), name='bh') # hidden bias

def _initial_W(self):
    return np.asarray( np.random.uniform(
        low = -4*np.sqrt(6./(self.n_hidden+self.
        n_visible)),
        high = 4*np.sqrt(6./(self.n_hidden+self.
        n_visible)),
        size = (self.n_visible, self.n_hidden)),
        dtype = theano.config.floatX)
```

```
    def _initial_bias(self, n):
        return np.zeros(n, dtype = theano.config.floatX)


class LearntPrecisionSeparateGaussianBinaryRBM(RBM):
    """
    Important: Wp and bvp should be constrained to be
    negative.
    This RBM models mean and precision with separate
    hidden units.
    """
    def __init__(self, n_visible, n_hidden_mean, n_
hidden_precision):
        super(LearntPrecisionSeparateGaussianBinaryRBM,
        self).__init__()
        # data shape
        self.n_visible = n_visible
        self.n_hidden_mean = n_hidden_mean
        self.n_hidden_precision = n_hidden_precision
        # units
        self.v = units.LearntPrecisionGaussianUnits
        (self, name='v') # visibles
        self.hm = units.BinaryUnits(self, name='hm')
        # hiddens for mean
        self.hp = units.BinaryUnits(self, name='hp')
        # hiddens for precision
        # parameters
        self.Wm = parameters.ProdParameters(self,
        [self.v, self.hm], theano.shared(value =
        self._initial_W(self.n_visible, self.n_
        hidden_mean), name='Wm'), name='Wm') # weights
        self.Wp = parameters.ProdParameters(self,
        [self.v.precision_units, self.hp], theano.
        shared(value = -np.abs(self._initial_W(self.
        n_visible, self.n_hidden_precision))/1000,
        name='Wp'), name='Wp') # weights
        self.bvm = parameters.BiasParameters(self,
        self.v, theano.shared(value = self._initial_
```

```python
        bias(self.n_visible), name='bvm'), name='bvm')
        # visible bias
        self.bvp = parameters.BiasParameters(self,
        self.v.precision_units, theano.shared(value =
        self._initial_bias(self.n_visible),
        name='bvp'),
        name='bvp') # precision bias
        self.bhm = parameters.BiasParameters(self,
        self.hm, theano.shared(value = self._initial
        _bias(self.n_hidden_mean), name='bhm'), name=
        'bhm') # hidden bias for mean
        self.bhp = parameters.BiasParameters(self,
        self.hp, theano.shared(value = self._initial_
        bias(self.n_hidden_precision)+1.0, name='bhp'),
        name='bhp') # hidden bias for precision

    def _initial_W(self, nv, nh):
        return np.asarray( np.random.uniform(
                low  = -4*np.sqrt(6./(nv+nh)),
                high = 4*np.sqrt(6./(nv+nh)),
                size = (nv, nh)),
                dtype = theano.config.floatX)

    def _initial_bias(self, n):
        return np.zeros(n, dtype = theano.config.floatX)

class BinaryBinaryCRBM1(BinaryBinaryRBM):
    def __init__(self, n_visible, n_hidden,
    n_context):
        super(BinaryBinaryCRBM1, self).__init__
        (n_visible, n_hidden)
        # data shape
        self.n_context = n_context
        # units
        #self.x = units.Units(self, name='x') # context
        self.x = units.SoftmaxUnits(self, name='s')
        # parameters
```

```
    self.pmap['A'] = theano.shared(value = self._
    initial_A(), name='A')
    self.pmap['B'] = theano.shared(value = self._
    initial_B(), name='B')
    self.A = parameters.ProdParameters(self,
    [self.x, self.v], 'A', name='A') # context-to-
    visible weights
    self.B = parameters.ProdParameters(self,
    [self.x, self.h], 'B', name='B') # context-to-
    hidden weights

  def _initial_A(self):
    return np.zeros((self.n_context, self.
    n_visible), dtype = theano.config.floatX)

  def _initial_B(self):
    return np.zeros((self.n_context, self.
    n_hidden), dtype = theano.config.floatX)


class TruncExpBinaryRBM(RBM): # RBM with truncated
exponential visibles and binary hiddens
  def __init__(self, n_visible, n_hidden):
    super(TruncExpBinaryRBM, self).__init__()
    # data shape
    self.n_visible = n_visible
    self.n_hidden = n_hidden
    # units
    self.v = units.TruncatedExponentialUnits(self,
    name='v') # visibles
    self.h = units.BinaryUnits(self, name='h')
    # hiddens
    # parameters
    self.W = parameters.ProdParameters(self,
    [self.v, self.h], theano.shared(value = self._
    initial_W(), name='W'), name='W') # weights
    self.bv = parameters.BiasParameters(self,
```

```
        self.v, theano.shared(value = self._initial_
        bv(), name='bv'), name='bv') # visible bias
        self.bh = parameters.BiasParameters(self,
        self.h, theano.shared(value = self._initial_
        bh(), name='bh'), name='bh') # hidden bias

    def _initial_W(self):
#       return np.asarray( np.random.uniform(
#               low  = -4*np.sqrt(6./(self.n_hidden+self.
                n_visible)),
#               high = 4*np.sqrt(6./(self.n_hidden+self.
                n_visible)),
#                size = (self.n_visible, self.n_hidden)),
#                 dtype = theano.config.floatX)

        return np.asarray( np.random.normal(0, 0.01,
                size = (self.n_visible, self.n_hidden)),
                dtype = theano.config.floatX)

    def _initial_bv(self):
        return np.zeros(self.n_visible, dtype = theano.
        config.floatX)

    def _initial_bh(self):
        return np.zeros(self.n_hidden, dtype = theano.
        config.floatX)
```

3. prediction.py

```
from collections import OrderedDict
import theano
from theano import tensor as T
import numpy as np

from morb import parameters

def label_prediction(rbm, vmap, pmap, visible_units,
label_units, hidden_units, context_units=[], name=
```

```
'func', mb_size=32, mode=None, logprob=True, only_
activation=False):
  """ Calculate p(y|v), the probability of the labels
given the visible state.
  $
    p\left(y\left|v\right.\right) = \frac{
    \exp\left( b_y + \sum_j \mathrm{softplus} \left
( c_j + U_{jy} + \sum_i W_{ji} x_i \right) \right)
    }{
        \sum_{y^*} \exp\left( b_y + \sum_j \mathrm
{softplus} \left( c_j + U_{jy^*} + \sum_i W_{ji} x_i
\right) \right)
    }
  $

  Based on:
  [1] Larochelle, H., Mandel, M., Pascanu, R., &
      Bengio, Y. (2012).
      Learning Algorithms for the Classification
      Restricted Boltzmann Machine.
      Journal of Machine Learning Research, 13,
      643-669.

  :param rbm:
    the RBM
  :type rbm:
    morb.base.RBM
  :param vmap:
    dictionary with RBM value formulas
  :param visible_units:
    the visible units (v)
  :type visible_units:
    list of morb.base.Units
  :param label_units:
    the label units (y)
  :type label_units:
    list of morb.base.Units
  :param hidden_units:
```

```
      the hidden units (h)
 :type visible_units:
      list of morb.base.Units
 """

probability_map = []

# Larochelle, 2012
for y in label_units:
    all_params_y = rbm.params_affecting(y)

    # bias for labels
    by = [ param for param in all_params_y if param.
    affects_only(y) ]
    assert len(by)==1
    assert isinstance(by[0], parameters.
    BiasParameters)

    # (minibatches, labels)

    by_weights_for_v = T.shape_padleft(pmap[by[0].
    var], 1)

    # collect all components

    label_activation = by_weights_for_v

    for h in hidden_units:
        h_act_given_v = h.activation(vmap, pmap,
        skip_units=label_units)

        # weights U connecting labels to hidden
        U = [ param for param in all_params_y if param.
        affects(h) ]

        assert len(U)==1
        assert U[0].weights_for
```

```
    # sum over hiddens
    a = T.nnet.softplus(U[0].weights_for(y, pmap)
    + T.shape_padright(h_act_given_v, 1))

    # sum over hiddens
    a = T.sum(a, axis=range(1,a.ndim - 1))
    # result: (minibatches, labels)
    label_activation += a

if only_activation:
  probability_map.append(label_activation)

elif not logprob:
  label_activation = T.exp(label_activation)

  # normalise over labels
  label_activation = label_activation / T.sum
  (label_activation, axis=1, keepdims=True)

  # (minibatches, labels)
  probability_map.append(label_activation)

else:
 # for numerical stability (no exp of large numbers)
 # see http://lingpipe-blog.com/2009/06/25/log-
 sum-of-exponentials/
 max_label_activation = T.max(label_activation,
 axis=1, keepdims=True)
 normalised_label_activation = \
     label_activation \
     - max_label_activation \
     - T.log(1e-20 + T.sum(T.exp(label_activation -
       max_label_activation), axis=1, keepdims=
       True))

  # (minibatches, labels)
  probability_map.append(normalised_label_
  activation)
```

```
# initialise datasets
data_sets = OrderedDict()
for u in visible_units:
   shape = (1,) * vmap[u].ndim
   data_sets[u] = theano.shared(value = np.zeros
   (shape, dtype=theano.config.floatX),
                 name="dataset for '%s'" % u.name)

for u in context_units:
   shape = (1,) * vmap[u].ndim
   data_sets[u] = theano.shared(value = np.zeros
   (shape, dtype=theano.config.floatX),
                 name="dataset for '%s'" % u.name)

index = T.lscalar() # index to a minibatch

# construct givens for the compiled theano function -
mapping variables to data
givens = dict((vmap[u], data_sets[u][index*
mb_size:(index+1)*mb_size]) for u in list
(visible_units)+list(context_units))

  TF = theano.function([index], probability_map,
  givens = givens, name = name, mode = mode)
#  theano.printing.debugprint(hidden_units[0].
activation(vmap))
#  vmap[hidden_units[0]] = hidden_units[0].
activation(vmap)
#  theano.printing.debugprint(label_units[0].
activation(vmap))
#  theano.printing.debugprint(TF)

 def func(dmap):
   # dmap is a dict that maps unit types on their
   respective datasets (numeric).
   units_list = dmap.keys()
   data_sizes = [int(np.ceil(dmap[u].shape[0] /
   float(mb_size))) for u in units_list]
```

```
    if data_sizes.count(data_sizes[0]) != len
    (data_sizes): # check if all data sizes are equal
        raise RuntimeError("The sizes of the supplied
        datasets for the different input units are not
        equal.")

  data_cast = [dmap[u].astype(theano.config.floatX)
  for u in units_list]

   for i, u in enumerate(units_list):
      data_sets[u].set_value(data_cast[i],
      borrow=True)

   for batch_index in xrange(min(data_sizes)):
      yield TF(batch_index)

return func
```

Research on insect pest image detection and recognition based on bio-inspired methods

8.1 INTRODUCTION

The occurrence of insect pests can have significant negative effects on the quality and quantity of agricultural products. If the pests are not detected in time, there will be an increase in food insecurity[1]. Pests are usually detected manually by agriculture experts. This task requires continuous monitoring of crops and is subjective, labour intensive, and expensive for large farms[2]. The rapid development of image processing technology has provided a new way for pest recognition, which can not only greatly improve the recognition efficiency, but also solve problems such as lack of agriculture experts and poor objectivity[3].

In recent years, lots of research has been made on pest detection and recognition based on image processing technology. Gassoumi et al.[4] used a neural network-based approach for insect classification in cotton ecosystems and achieved an accuracy of 90%. Li et al.[5] proposed a new pest detection method based on stereo vision to get the location information of the pest. Faithpraise and Chatwin[1] proposed an automatic

DOI: 10.1201/9781003281641-8

method for pest detection and recognition using k-means clustering algorithm and correspondence filters. Boissard et al.[6] presented a cognitive vision system to detect white flies in greenhouse crops and demonstrated that automatic processing was reliable. Dey et al.[7] presented an automated approach for detecting white fly pests from leaf images and used the k-means clustering method to segment pests from infected leaves. Roldan-Serrato et al.[8] developed a recognition system based on a special neural network - the random subspace classifier - for the Colorado potato beetle and obtained a recognition rate of 85%. Wen et al.[9] proposed the IpSDAE architecture to build a deep neural network for moth identification and achieved a good identification accuracy of 96.9%. Cheng et al.[10] proposed a pest identification method using deep residual learning and achieved an accuracy of 98.67% for 10 classes.

Despite great achievements in pest recognition in recent years, most research was not performed in outdoor field environments and was carried in highly controlled lab environments instead of using real-world scenes. However, in actual field conditions, complex environments, different viewpoints, and poses impose great challenges for pest detection and recognition. Though some research involved automatic detection in natural environments, most concerned only one species or had high time complexity. Detecting pests rapidly and accurately and extracting features that are invariant to viewpoint, scale, and lighting conditions are crucial for the recognition of crop pests.

Humans can recognize objects rapidly, no matter how complex the conditions are. This outstanding ability is supported by the visual system. Inspired by the research findings of cognitive neuroscience, some computational models have been proposed in recent years to model the human visual system. A well-known model, HMAX[11] showed outstanding performance in object recognition tasks. Although HMAX shows good invariance to position and scale, it is sensitive to rotation. So this chapter extended the HMAX by integrating Scale Invariant Feature Transform (SIFT)[12] and Non-negative Sparse Coding (NNSC) into it, which was denoted as the SIFT-HMAX model. First, the Saliency Using Natural statistics (SUN) model was used to detect the pest and extract ROIs. Then the SIFT-HMAX model and LCP algorithm were used to extract the invariant features. Finally, the extracted features were standardized and fed to a Support Vector Machine to perform recognition.

8.2 MATERIALS AND METHODS

8.2.1 Materials

We investigated a collection of ten categories of insect pests (mainly affecting tea plants), which are: Locusta migratoria, Parasa lepida, Euproctis pseudoconspersa Strand, Empoasca flavescens, Spodoptera exigua, Chrysochus chinensis, larva of Laspeyresia pomonella, larva of Spodoptera exigua, Acrida cinerea, and Laspeyresia pomonella. Some of the sample images are presented in Fig. 8.1. Each category contained about 40–70 sample images. Among these samples, some were collected from online resources (see Appendix), such as Insert Images, IPM Images, Dave's Garden, and so on. The others were taken outdoors using a digital SLR camera, which have been uploaded to Mendeley Data, and the links are provided in the Appendix. The sample images show great variation in scale, position, viewpoint, lighting conditions, and backgrounds.

8.2.2 Methods

The complete flowchart of our method is illustrated in Fig. 8.2. First SUN model was used to create the saliency map and extract the ROIs. Then, invariant features representing the pests were extracted using the SIFT-HMAX model and LCP algorithm. Finally, the extracted features were fed into a Support Vector Machine to perform the recognition task.

8.2.2.1 Object detection

In order to detect the object rapidly and accurately, the saliency map was used to target the object of interest. Saliency map is an effective and biologically feasible method for modelling visual attention. A typical way of generating a saliency map is to assign high saliency to regions with rare features. One way to represent rare features is to determine how frequently they occur. By fitting a distribution $P(F)$, where F denotes image features, rare features can be immediately obtained by computing $P(F)^{-1}$.

First, image patches were extracted from random locations of images in McGill colour image dataset[13]. Next, Principal Component Analysis (PCA) was used on these image patches to reduce dimensionality. Then, efficient FastICA was used to the image patches and obtained ICA filters,

(a)

(b)

(c)

(d)

(e)

(f)

(g)

(h)

(i)

(j)

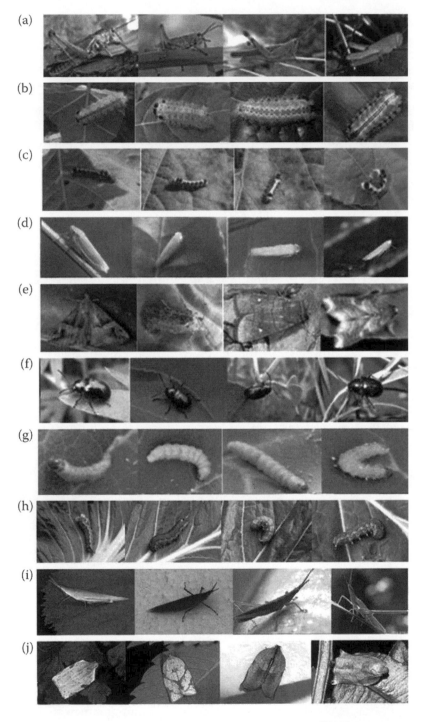

(Caption on next page)

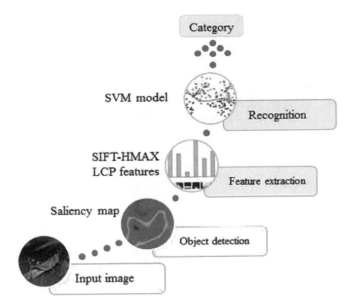

FIGURE 8.2   Illustration of overall flowchart.

as shown in Fig. 8.3. When ICA filters were applied to image patches, they produced similar properties to V1 neurons[14].

Then, ICA features were extracted by applying ICA filters on the input images. As these features are largely statistically independent, $P(F)$ can be defined as the product of unidimensional distributions:

$$P(F = f) = \prod_{i} P(f_i) \qquad (8.1)$$

where f_i is the ith element of vector f. Linear distribution $P(f_i)$ can be formulated by a generated Gaussian distribution[14]:

$$P(f_i) = \frac{\phi_i}{2s_i \Gamma(\phi_i^{-1})} \exp\left(-\left|\frac{f_i}{s_i}\right|^{\phi_i}\right) \qquad (8.2)$$

FIGURE 8.1   Sample images taken in natural conditions. (a) Locusta migratoria; (b) Parasa lepida; (c) Gypsy moth larva; (d) Empoasca flavescens; (e) Spodoptera exigua; (f) Chrysochus chinensis; (g) Laspeyresia pomonella larva; (h) Spodoptera exigua larva; (i) Atractomorpha sinensis; (j) Laspeyresia pomonella.

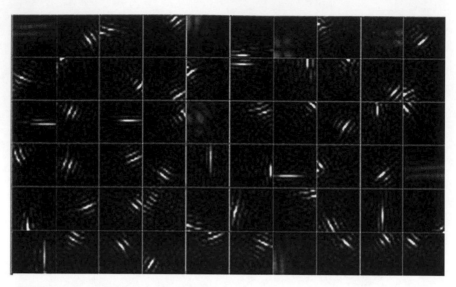

FIGURE 8.3   Some learned ICA filters.

where ϕ_i and s_i indicate the shape and scale parameters which can be estimated by a convex function, and Γ denotes the Gamma function.

Thus the saliency map of the input image can be obtained. To facilitate the subsequent processing, the saliency map was normalized to the range of 0 to 1. Then, the object was separated from the background using an adaptive threshold that was computed using Otsu's method. If the saliency of a pixel is larger than the threshold, the pixel value was set to 1; otherwise, it was set to 0. This would result in a binary image, on which the largest connected region usually corresponded to the object of interest. Then we chose the minimum bounding rectangle of the largest connected region as the ROI. Fig. 8.4 demonstrates the saliency maps and the ROIs of some image samples. The first row shows raw images, the second row their corresponding saliency maps, and the third row the corresponding ROIs. Though there was great variation in the colour and backgrounds of the sample images, our method could accurately target the object and extract the ROIs from the complex backgrounds.

8.2.2.2 SIFT-HMAX model
The standard HMAX is a bio-inspired model with four layers of computational units: S1, C1, S2, and C2. The proposed SIFT-HMAX

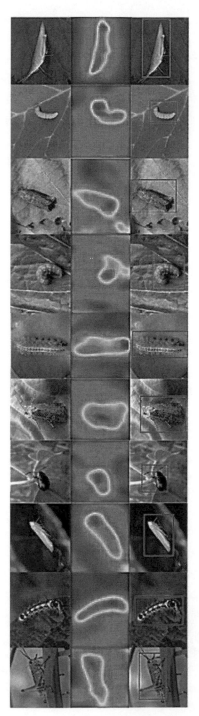

FIGURE 8.4   Ground truth, the corresponding saliency maps, and ROIs.

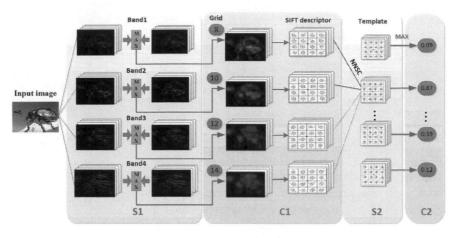

FIGURE 8.5   The architecture of the SIFT-HMAX model.

model consists of two contributions to the HMAX model in C1 and S2 layers, as depicted in Fig. 8.5.

The ROI (colour image) obtained above was first converted into grey images by:

$$I = 0.3R + 0.59G + 0.11B \tag{8.3}$$

where R, G, and B are respectively the red, green, and blue channels of the colour image.

S1 layer received grey-value images as input and applies a set of Gabor filters to detect edge. The function of Gabor filter for pixel position (x, y) is given by:

$$G(x, y) = \exp\left(-\frac{x_0^2 + \gamma^2 y_0^2}{2\sigma^2}\right)\cos\left(\frac{2\pi}{\lambda}x_0\right), \text{ s. t.}$$
$$x_0 = x \cos\theta + y \sin\theta \text{ and } y_0 = -x \sin\theta + y \cos\theta \tag{8.4}$$

In this equation, x and y are position coordinates; λ, θ, σ, and γ are filter parameters, which respectively represent the wavelength, orientation, standard deviation, as well as the spatial aspect ratio of the Gabor filter.

According to the research findings of Eliasi et al.[15], not all bands are necessary to achieve better performance and the first four bands of Gabor filters are more efficient than using all bands. To reduce the processing time, only the first four bands in S1 layer were used. Fig. 8.6

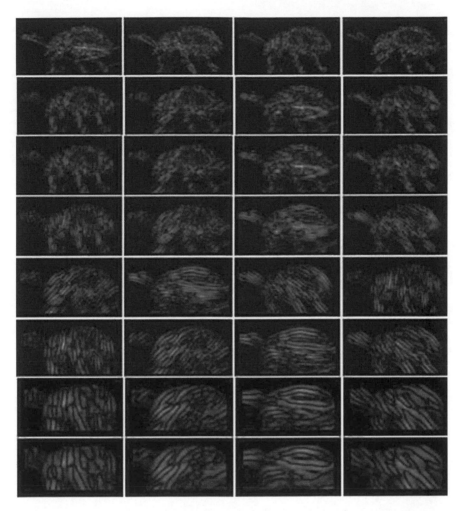

FIGURE 8.6 Filtering results of Gabor filters at 8 scales and 4 orientations.

demonstrates the filtering results of Gabor filters at 4 orientations and 8 scales.

C1 units first pooled over afferent S1 units from the same scale band and orientation (as shown in Fig. 8.7). Next, SIFT descriptors were extracted from the subsampled images with dense SIFT method. Each image was densely divided into $K \times K$ sub-patches with a certain step, and then a SIFT descriptor was extracted on each sub-patch, which resulted in a 128-dimensional vector.

Considering the sparse response properties of V4 neurons, NNSC was used to compute the S2 unit response instead of Euclidean distance[16].

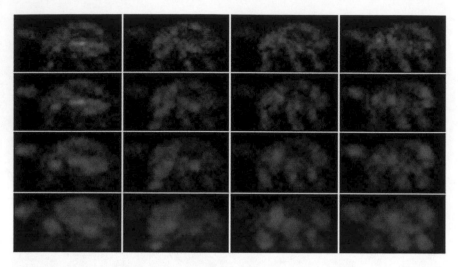

FIGURE 8.7   Results of local pooling.

Suppose X is the set of C1 image patches and T is the set of image templates, then X can be linearly presented by:

$$X = TC = \begin{bmatrix} t_{11} & t_{12} & \cdots & t_{1N} \\ t_{21} & t_{22} & \cdots & t_{2N} \\ \vdots & \vdots & \vdots & \vdots \\ t_{D1} & t_{D2} & \cdots & t_{DN} \end{bmatrix} \cdot \begin{bmatrix} c_{11} & c_{12} & \cdots & c_{1M} \\ c_{21} & c_{22} & \cdots & c_{2M} \\ \vdots & \vdots & \vdots & \vdots \\ c_{N1} & c_{N2} & \cdots & c_{NM} \end{bmatrix} \quad (8.5)$$

where C is the coefficient of non-negative sparse coding which can be computed by solving the following optimization problem:

$$C = \arg\min_{S} \left(\|X - TS\|_F^2 + \alpha\|S\|_1 \right) \quad (8.6)$$

where α is a regularization parameter controlling the tradeoff between sparseness and accurate reconstruction.

The C2 unit response can be achieved by pooling S2 response matrix C with column-wise maximum and results in an N-dimensional vector:

$$R = \max C = \left(\max_j c_{1j}, \ \max_j c_{2j}, \ \cdots, \max_j c_{Nj} \right) \quad (8.7)$$

where N is the number of image templates.

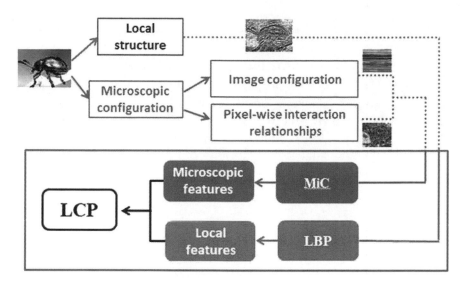

FIGURE 8.8   The feature extraction framework of LCP.

8.2.2.3 Extraction of LCP features

Image textures reflect the information of spatial distributions, grey level statistics, and so on. The Local Binary Pattern (LBP) descriptor is widely used because of its robustness to lighting and rotational changes, and its high computational efficiency[17]. However, it lacks the information of microscopic pixel-wise. LCP features capture pixel-wise interaction information and produce discriminative representation combined with LBP[18].

The information architecture of LCP consists of two levels: local structural information and microscopic configuration information, as shown in Fig. 8.8. LBP is used to obtain local structure, and a microscopic configuration (MiC) model is used to encode microscopic information. For an input image, an 81-dimensional vector can be obtained.

8.3   EXPERIMENTAL RESULTS

For each category, 20 samples were selected randomly as training set, and the rest were used for testing. First, the ROI was extracted from the input image, and then invariant features were extracted from the ROI. Finally, the features were normalized to mean 0 and variance 1 and fed into linear support vector machine (libSVM[19]) to perform classification task. The whole process was repeated five times and the mean was taken to eradicate the discrepancies. The proposed model and contrast models were all performed under Matlab R2015b platform. All experiments were conducted on a laptop with an Intel Core i5-4200 CPU 2.3 GHz and 4G RAM.

8.3.1 Effectiveness analysis of template number

In the SIFT-HMAX model, the number of image templates is an important parameter. To evaluate its effectiveness on the recognition performance, we selected six categories (Locusta migratoria, Parasa lepida, Gypsy moth larva, Empoasca flavescens, Spodoptera exigua, and Chrysochus chinensis) as positive datasets and background as negative datasets. Fig. 8.9 shows the recognition performance with different number of image templates (10, 50, 100, 200, and 500). In general, as for SIFT-HMAX itself, the more templates there are, the better performance can be achieved. When there are more than 100 templates, a reasonable result can be obtained. Given that LCP feature is an 81-dimension vector, too many templates will weaken the effectiveness of LCP features. As a compromise, we used 100 templates in SIFT-HMAX model.

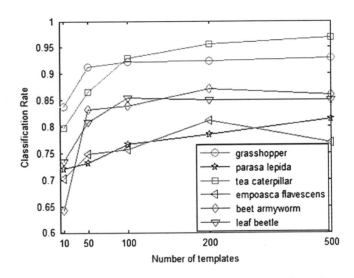

FIGURE 8.9 Classification performance with different number of templates.

8.3.2 Results of object detection

To demonstrate the effectiveness of the detection method, we compared it with several saliency methods: AIM[20], GBVS[21], and SIM[22], as shown in Fig. 8.10. Generally speaking, all methods can detect the object accurately in the presence of simple backgrounds (such as the second and third rows). But for complex situations (such as the first and fifth rows), the compared methods – especially AIM and SIM methods – could not perfectly extract the ROIs. However, our method could accurately detect the object in all cases, no matter how complex the conditions were.

8.3.3 Performance evaluation

The extracted features were fed to SVM to perform classification task. SVM model used RBF kernel function, and the optimized value of parameter C was 3.03 and g was 0.068. Fig. 8.11 illustrates the confusion matrix of the proposed method.

Relatively speaking, Parasa lepida, Empoasca flavescens, and Laspeyresia pomonella larva have achieved excellent recognition results with an accuracy of 100%. However, the recognition rates of Locusta migratoria, Laspeyresia pomonella, and Spodoptera exigua are relatively low and less than 80%, which is partly because there is a big intra-class variance in the morphology, texture, and background, such as Locusta migratoria. Meanwhile, some features are similar across species, such as Laspeyresia pomonella and Spodoptera exigua. So, considering both intra-class diversity and extra-class distinctiveness at the same time is a tough and challenging task. Moreover, the samples come from different sources and websites, which results in large variation in image resolution and lighting conditions. In addition, we use the same parameters for all images, such as image patch size and Gabor filter parameters, which may lead to misclassification, to some extent.

To evaluate the recognition performance of our method, we compared with some prevalent methods (HMAX[11], Sparse coding[23], NIMBLE[14], and a Deep Convolutional Network (vgg19[24]) on both recognition accuracy and processing time, as illustrated in Table 8.1. There are 19 layers with learnable weights: 16 convolutional layers and 3 fully connected layers in the network. The architecture and configurations of the model were as detailed in the research of Simonyan and Zisserman[24].

Source image	AIM	GBVS	SIM	Our method

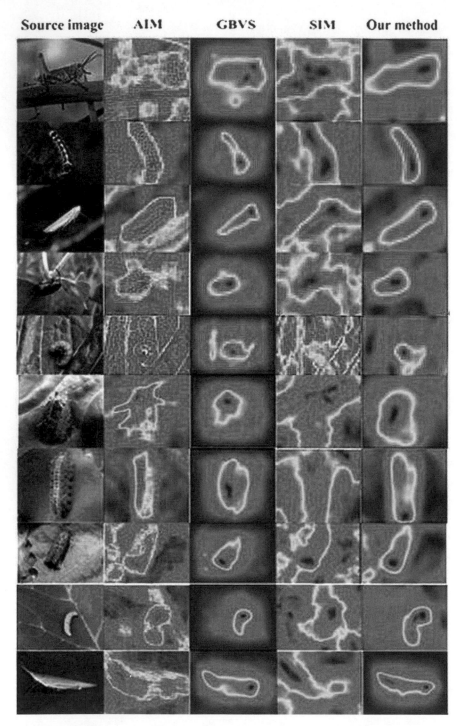

(Caption on next page)

	Locusta migratoria	Laspeyresia pomonella	Parasa lepida	Euproctis pseudocons persa Strand	Empoasca flavescens	Spodoptera exigua	Chrysochus chinensis	Laspeyresia pomonella larva	Spodoptera exigua larva	Acrida cinerea
Locusta migratoria	78.26	0.00	0.00	0.00	0.00	6.25	10.00	0.00	0.00	0.00
Laspeyresia pomonella	0.00	76.92	0.00	0.00	0.00	18.75	0.00	0.00	0.00	0.00
Parasa lepida	0.00	15.38	100.00	16.67	0.00	0.00	0.00	0.00	0.00	0.00
Euproctis pseudoconspersa Strand	4.35	0.00	0.00	83.33	0.00	0.00	0.00	0.00	0.00	6.67
Empoasca flavescens	0.00	0.00	0.00	0.00	100.00	0.00	0.00	0.00	0.00	0.00
Spodoptera exigua	13.04	7.69	0.00	0.00	0.00	75.00	0.00	0.00	0.00	6.67
Chrysochus chinensis	4.35	0.00	0.00	0.00	0.00	0.00	90.00	0.00	0.00	0.00
Laspeyresia pomonella larva	0.00	0.00	0.00	0.00	0.00	0.00	0.00	100.00	7.14	6.67
Spodoptera exigua larva	0.00	0.00	0.00	0.00	0.00	0.00	0.00	0.00	92.86	0.00
Acrida cinerea	0.00	0.00	0.00	0.00	0.00	0.00	0.00	0.00	0.00	80.00

FIGURE 8.11   Confusion matrix of the proposed method.

TABLE 8.1   Comparison of different methods

Methods	Accuracy (%)	Time(S)
Our method	85.5	146.3
HMAX	69.2	476.3
Sparse coding	79.3	251.4
NIMBLE	83.1	684.7
Vgg19	86.9	345.5

As there are millions of free parameters to be optimized through an extensive training phase in MatConvNet, to save time, we used a pre-trained model (imagenet-vgg-verydeep-19) which is available at http://www.vlfeat.org/matconvnet/pretrained/. So the processing time of MatConvNet only involves time spent on feature extraction and does not include time on model training.

Experimental results demonstrate that the proposed method has significantly outperformed the original HMAX model both in recognition rate and processing time. With respect to the recognition rate, the proposed method performed better than the contrast methods except for

FIGURE 8.10   Comparison among several saliency methods.

MatConvNet. Though MatConvNet achieved slightly higher recognition accuracy than our method, it is a large neural network with millions of free parameters optimized through extensive training on large training datasets. Even using a pre-trained model, MatConvNet still takes more time than our method under the same operating environment. Our method has great advantage on processing time and achieved the best time efficiency. In general, the proposed method is appropriate for rapid detection and recognition of insect pests under complex and natural conditions.

8.4 DISCUSSION

Object detection is critical for pest recognition, which is traditionally achieved by image segmentation. In reality, images of insect pests often have a complicated background, which makes it very difficult to separate them from their background. Segmentation algorithms for complicated backgrounds are very complex and have poor practicability and operability. Most research was performed under strictly controlled conditions and focused on simple backgrounds[25]. So the existing methods are not very suitable for the detection of insect pests under natural conditions. Inspired by the mechanism of human visual attention, this chapter applied saliency maps for pest detection and achieved satisfying results.

According to the appearance characteristics of insect pests, this chapter attempted to apply biological visual perception mechanisms to pest recognition. SIFT-HMAX model could extract features which were invariant to changes in position, scale, and rotation. LCP features were robust to rotational and lighting changes. Combining both together can well describe the invariant features representing the shape and texture of insect pests. Though there are some parameters to be tuned in the SIFT-HMAX model, related research[26] has provided references and methods for selecting these parameters. So there is no need for complex parameter tuning.

Limited by the number of samples, this chapter mainly focuses on recognition methods with small training samples. The insect pests occurred infrequently, which brought great difficulties for data collection, especially in natural environments. Meanwhile, there have not been enough public insect pest image datasets available until now. Though Wang et al.[27] have created a sharing data platform for insect image research. However, most insect images are taken in a controlled lab environment, which can't well satisfy the requirement of insect recognition in field conditions. As to our images, some were taken outdoors using a digital camera, and others were collected from various websites. As a result, there was great variation in image dimensions and shooting environments, which brought great challenges to insect recognition. But on the other hand, such images are more representative of actual conditions. To improve the recognition performance, the number of image samples need to be enlarged further, and the pest dataset should also be standardized further and form a public dataset for convenience of access.

8.5 CONCLUSIONS

Inspired by human visual attention and perception mechanisms, we first detected the pests and extracted the ROIs using a natural statistics model. Then we proposed an invariant feature extraction method of insect pests based on the SIFT-HMAX model and LCP features. And, finally, the SVM model was used to perform the recognition task. Experimental results showed that our detection method could exactly target and detect the object region in images from complex natural environments. The proposed method demonstrated good recognition performance with an accuracy of 85.5% and provides a new method for rapid detection and recognition of insect pests.

REFERENCES

1. Fina, F., Birch, P., Young, R., Obu, J., Faithpraise, B., and Chatwin, C. 2013. Automatic plant pest detection and recognition using k-means clustering algorithm and correspondence filters. *International Journal of Advanced Biotechnology and Research* 4(2):189–199.
2. Al-Hiary, H., Bani-Ahmad, S., Reyalat, M., Braik, M., and Alrahamneh, Z. 2011. Fast and accurate detection and classification of plant diseases. *International Journal of Computer Applications* 17(1):31–38.
3. Hu, Y. Q., Song, L. T., Zhang, J., Xie, C. J., and Li, R. 2014. Pest image recognition of multi-feature fusion based on sparse representation. *Pattern Recognition and Artificial Intelligence* 27(11):985–992.
4. Gassoumi, H., Prasad, N. R., and Ellington, J. J. 2000. Neural network-based approach for insect classification in cotton ecosystems. *In International Conference on Intelligent Technologies.*
5. Li, Y., Xia, C., and Lee, J. 2009, July. Vision-based pest detection and automatic spray of greenhouse plant. *In 2009 IEEE international symposium on industrial electronics*: 920–925.

6. Boissard, Paul, Vincent Martin, and Sabine Moisan. 2008. A cognitive vision approach to early pest detection in greenhouse crops. *Computers and Electronics in Agriculture* 62(2):81–93. 10.1016/j.compag.2007.11.009.

7. Dey, A., D. Bhoumik, and K. N. Dey. Automatic detection of whitefly pest using statistical feature extraction and image classification methods. *International Research Journal of Engineering and Technology* 3(9): 950–959.

8. Roldán-Serrato, L., Baydyk, T., Kussul, E., Escalante-Estrada, A., and Rodriguez, M. T. G. 2015, June. Recognition of pests on crops with a random subspace classifier. *In 2015 4th International Work Conference on Bioinspired Intelligence (IWOBI)*:21–26.

9. Wen, C., Wu, D., Hu, H., and Pan, W. 2015. Pose estimation-dependent identification method for field moth images using deep learning architecture. *Biosystems Engineering* 136:117–128.

10. Cheng, X., G. Geng, M. Zhou, and S. Huang. 2009. Applying expectation-maximization in insect image segmentation using multi-features. *Computer Applications and Software* 26(2):20–22.

11. Serre, T., Wolf, L., Bileschi, S., Riesenhuber, M., and Poggio, T. 2007. Robust object recognition with cortex-like mechanisms. *IEEE Transactions on Pattern Analysis and Machine Intelligence* 29(3):411–426.

12. Lowe, D. G. 2004. Distinctive image features from scale-invariant keypoints. *International Journal of Computer Vision* 60(2):91–110.

13. Olmos, A., and Kingdom, F. A. 2004. A biologically inspired algorithm for the recovery of shading and reflectance images. *Perception* 33(12): 1463–1473.

14. Kanan, C., and Cottrell, G. 2010, June. Robust classification of objects, faces, and flowers using natural image statistics. In *2010 IEEE Computer Society Conference on Computer Vision and Pattern Recognition*: 2472–2479.

15. Eliasi, M., Yaghoubi, Z., and Eliasi, A. 2011. Intermediate layer optimization of HMAX model for face recognition. *In 2011 IEEE International Conference on Computer Applications and Industrial Electronics (ICCAIE)*: 432–436.

16. Wang, Y., and Deng, L. 2016. Modelling object recognition in visual cortex using multiple firing k-means and non-negative sparse coding. *Signal Processing* 124:198–209.

17. Brahnam, S., Lakhmi C., Jain, Alessandra, Lumini, and Loris Nanni. 2014. Introduction to local binary patterns: New variants and applications. *Studies in Computational Intelligence.* 10.1007/978-3-642-39289-4_1.

18. Guo, Y., Zhao, G., and Pietikäinen, M. 2011, August. Texture classification using a linear configuration model based descriptor. *In BMVC*:1–10.

19. Chang, C., and Lin, C. 2011. LIBSVM: A Library for support vector machines. *ACM Transactions on Intelligent Systems and Technology (TIST)* 2:1–39. 10.1145/1961189.1961199.

20. Bruce, Neil D. B., and John K., Tsotsos. 2009. Saliency, attention, and visual search: An information theoretic approach. *Journal of Vision* 3(9):1–24. 10.1167/9.3.5.
21. Harel, J., Koch, C., and Perona, P. 2006. Graph-based visual saliency. *Conference on Advances in Neural Information Processing Systems.* Cambridge: MIT Press.
22. Murray, N., Vanrell, M., Otazu, X., and Parraga, C. A. 2011, June. Saliency estimation using a non-parametric low-level vision model. *CVPR* 2011:433–440.
23. Yang, J., Yu, K., Gong, Y., and Huang, T. 2009, June. Linear spatial pyramid matching using sparse coding for image classification. *In 2009 IEEE Conference on computer vision and pattern recognition*: 1794–1801.
24. Simonyan, K., and Zisserman, A. 2014. Very deep convolutional networks for large-scale image recognition. *arXiv preprint arXiv*:1409.1556.
25. Bodhe, Trupti S., and Prachi Mukherji. 2013. Selection of color space for image segmentation in pest detection. *In 2013 International Conference on Advances in Technology and Engineering*, ICATE 2013. 10.1109/ICAdTE.2013.6524753.
26. Meyers, E., and Wolf, L. 2008. Using biologically inspired features for face processing. *International Journal of Computer Vision* 76(1): 93–104.
27. Wang, J., Ji, L., Wang, L., Gao, L., and Shen, Z. 2011, October. A sharing data platform for insect image researches. *In 2011 4th International Conference on Biomedical Engineering and Informatics (BMEI)* 4: 2046–2048.

APPENDIX

Orthoptera Images. https://www.insectimages.org/browse/taxthumb.cfm?order=159
Grasshoppers. https://www.ipmimages.org/browse/subthumb.cfm?sub=256
Lepidoptera Images. https://www.insectimages.org/browse/taxthumb.cfm?order=131
https://www.ipmimages.org/browse/autthumb.cfm?aut=31253
Johnny N. Dell's Images. https://www.ipmimages.org/browse/autthumb.cfm?aut=72226
John C. French Sr.'s Images. https://www.ipmimages.org/browse/subthumb.cfm?sub=2520
beet armyworm Spodoptera exigua. http://davesgarden.com/guides/bf/

Dave's Garden, Bug Files.
http://gaga.biodiv.tw/new23/cp03_7.htm
http://gaga.biodiv.tw/new23/cp04_32.htm
http://gaga.biodiv.tw/new23/cp03_3.htm
http://gaga.biodiv.tw/new23/cp03_80.htm
http://gaga.biodiv.tw/new23/cp03_20.htm
http://gaga.biodiv.tw/new23/cp04_51.htm
http://www.zhibaochina.com/Diag4/n/Show.aspx?
id=3212
http://www.zhibaochina.com/Diag4/n/Show.aspx?
id=3211
http://www.zhibaochina.com/Diag4/n/Show.aspx?
id=3081
http://soso.nipic.com/?q=%E8%9D%97%E8%99%AB
http://tupu.3456.tv/shucai/chonghai
Plutarco Echegoyen, Bugwood.org
Whitney Cranshaw, Colorado State University,
Bugwood.org
Metin GULESCI, Bugwood.org
Ansel Oommen, Bugwood.org.
Karla Salp, Washington State Department of
Agriculture, Bugwood.org
Johnny N. Dell, Bugwood.org
Blake Layton, Mississippi State University,
Bugwood.org
John C. French Sr., Retired, Universities:
Auburn, GA, Clemson and U of MO, Bugwood.org
Daren Mueller, Iowa State University, Bugwood.org
David Cappaert, Bugwood.org
Sangmi Lee, Museum Collections: Orthoptera, USDA
APHIS ITP, Bugwood.org

## CHAPTER 9

Carrot defect detection and grading based on computer vision and deep learning

9.1 INTRODUCTION

The carrot is an important vegetable with high nutritional and medicinal values, which is cultivated widely in the world. Carrot quality processing is an essential operation to detect defects and grade them before entering the market, and quality inspection and grading sale can improve the quality and market competitiveness of carrots[1]. At present, carrot quality inspection and grading are mainly conducted by humans, which poses great problems such as low efficiency and accuracy, lack of consistency and uniformity[2], and greatly affect the quality and efficiency of carrot grading. So reliable and efficient quality processing methods[3] are urgently needed to decrease labour. As an emerging and ongoing technology, computer vision has provided an accurate, highly efficient and non-destructive method for grading agricultural products[4] and has become one of the most popular methods for external quality inspection of carrots.

Han et al.[5] proposed the extraction algorithm of key parameters affecting the appearance quality of carrots such as fibrous roots, green heads and cracking based on image processing methods. The authors[6] developed an automated carrot sorting system based on machine vision and proposed detection methods for three kinds of surface defects: curvature, fibrous roots, and surface cracks. Xie et al.[1] proposed a detection method of surface defects such as green shoulders, bending, fibrous roots, and surfaces cracked and broken based on machine vision.

Despite the successful application of traditional techniques in the field of carrot defect detection, these traditional methods are complex due to the requirement of several distinct steps[2], and have great limits on accuracy and flexibility, which greatly limits their applications under actual conditions[7]. To overcome these limits, deep learning[8] provides practical solutions to address some of the most challenging problems in recent years. Deep learning learns the image features and automatically extracts features and achieves great advantages over traditional image processing methods[9]. It has been successfully used to detect defects in peaches[10], cucumbers[11], root-trimmed garlics[12], and apples[13]. It has also been used in the field of carrot appearance quality inspection. Zhu et al.[14,15] used the deep learning model to identify carrot appearance quality and demonstrated the feasibility of deep learning in carrot appearance quality inspection. Although the performance of the deep learning method has been greatly improved, efficiency problems came with it, and the detection efficiency of these methods makes the online quick detection of defects in carrots challenging. To the authors' knowledge, no online defect detection system has been constructed based on deep learning.

In the research of postharvest quality grading, previous studies[2,12] mainly aimed to grade agricultural products into normal and defective ones, and no further grading of the normal ones. In the actual processing of carrot quality, defective carrots are first picked out by some workers, and then normal ones are further classified into different grades by size by other workers. However, current research mainly focused on external defect detection, and no research involved automatic carrot grading of normal carrots. Therefore, a fully automated system for external defect detection and grading is still not available. So, it is necessary to develop effective methods for online defect detection and carrot grading to make emerging processing technology successful and beneficial in the carrot industry[16].

9.2 MATERIAL AND METHODS

9.2.1 Materials

Washed carrots of different sizes and appearance were collected from Qingdao Youtian Agricultural Development Co., Ltd, which performs carrot quality inspection and grading manually. Carrots are first cleaned by a washing machine and then transferred to the front of workers in order for them to pick out defective ones, and normal carrots are further classified into different grades by size. Carrots with different sizes and shapes to cover normal and defective carrots (abnormity, cracks, fibrous roots, fractures, and bad spots) were selected as experimental samples, as shown in Fig. 9.1.

9.2.2 Image acquisition and dataset collection

The whole process of the system mainly included four steps: image acquisition, image preprocessing, defect detection, and carrot grading, as presented in Fig. 9.2.

During image acquisition, a linear camera with a frame-triggered mode was used to obtain images of carrots moving at a high speed (about 4 m/s). Images of three views were acquired from different sides of the carrot. The middle one was a real image which was used for surface defect detection and carrot grading; the others were mirror images and were used only for surface defect detection. To achieve brightness equalization, referring to the Ohta colour space, we used a new grayscale method calculated by Equation (9.1).

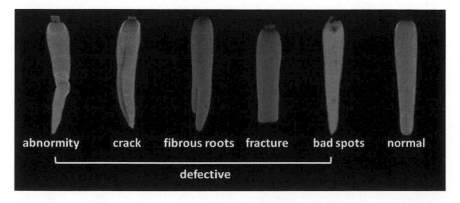

FIGURE 9.1 Normal carrots and defective carrots.

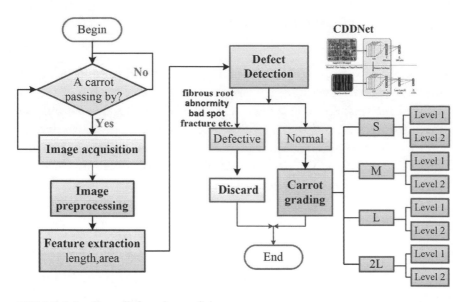

FIGURE 9.2 Overall flowchart of the system.

$$I_{gray} = 2.5I_r - 2I_g - 0.5I_b \tag{9.1}$$

where I_r, I_g, and I_b represent the red, green and blue components of the source image I, respectively. The process of image preprocessing is illustrated in Fig. 9.3.

Four image datasets were constructed (Table 9.1) by the image acquisition system for different experimental purposes, among which dataset 1 included two categories: normal carrots and defective carrots (abnormity, cracks, fibrous roots, fractures, and bad spots) and was used for parameter selection of CDDNet and performance evaluation of binary classification; dataset 2 was used for performance evaluation of muti-classification task; dataset 3 was for the MBR fitting; and dataset 4 was for convex polygon approximation.

9.2.3 Carrot defect detection model based on deep learning

Considering that ShuffleNet[17] is a very light network with very low complexity, and transfer learning has provided a very effective technology for object recognition, especially when there are limited training data[18], we constructed the carrot defect detection network (CDDNet) based on ShuffleNet and transfer learning to realize online defect detection of carrots, as illustrated in Fig. 9.4. The main innovation steps are as follows:

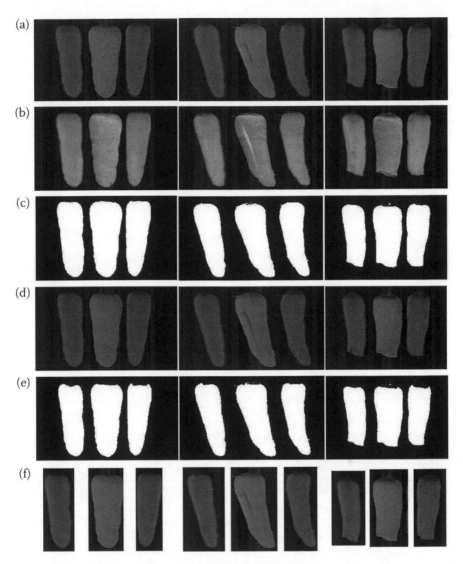

FIGURE 9.3 Illustration of image preprocessing. (a) Original carrot images of normal, cracked and broken carrots; (b) grey images calculated by Equation (9.1); (c) binary images of (b); (d) grey image by traditional method; (e) binary images of (d); and (f) individual images.

1. Remove the last ShuffleNet unit to reduce the model size;

2. Replace the average-pooling-based GlobalPool layer with a max-pooling layer;

TABLE 9.1   Description of four image datasets

Dataset Name	Categories	Training size	Validation size	Total size
Dataset 1	Normal	1709	733	2442
	Defective	3667	1571	5238
Dataset 2	Normal	700	300	1000
	Fibrous root	206	88	294
	Bad spot	411	76	487
	Abnormity	969	415	1384
Dataset 3	S	–	240	240
	M	–	246	246
	L	–	216	216
	2L	–	216	216
Dataset 4	Level 1	–	652	652
	Level 2	–	221	221

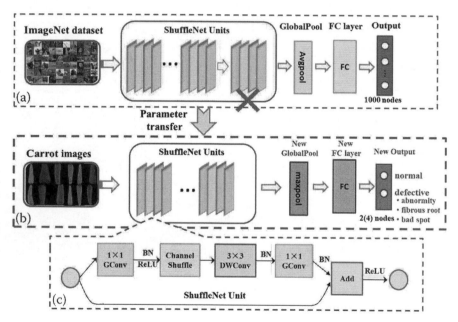

FIGURE 9.4   Model architecture of the proposed defect detection model. (a) Pre-trained ShuffleNet model; (b) CDDNet model; and (c) ShuffleNet unit.

3. Replace the fully connected (FC) layer and classification layer with a new FC layer and classification layer.

The overall architecture of CDDNet is presented in Table 9.2.

TABLE 9.2   Overall architecture of CDDNet

Layer	Output size	KSize	Stride	Repeat	Output channels
Image	224×224				3
Conv1	112×112	3×3	2		24
MaxPool	56×56	3×3	2		24
Stage2	28×28		2	1	136
	28×28		1	3	136
Stage3	14×14		2	1	272
	14×14		1	7	272
Stage4	7×7		2	1	544
	7×7		1	2	544
GlobalPool	1×1	7×7	1		
FC					1000
Output					2 (4)

9.2.4 Grading methods based on MBR fitting and convex polygon approximation

After defect detection, normal carrots were further classified into different specifications according to their length. Therefore, it is necessary to build the relationship between the pixel length and the actual length of the carrot. The pixel length can be measured by computing the minimum bounding rectangle of the carrot area. Therefore, a new regression method was proposed to predict the actual length from carrot images based on the minimum bounding rectangle approximation[19] and linear fitting[20], which can be described as follows:

Step 1 Convert the source image (Fig. 9.5a) into a binary image.

Step 2 Compute the minimum bounding rectangle of the carrot area (Fig. 9.5b) based on an approximation-based algorithm and MBR corner point positions tl (tl_x, tl_y), tr (tr_x, tr_y), bl (bl_x, bl_y), $br = (br_x, br_y)$ can be obtained using the formula in Equations (9.2)–(9.5):

$$tl_x = \frac{x_1 \tan \theta + x_3 \cot \theta + y_3 - y_1}{\tan \theta + \cot \theta},$$

$$tl_y = \frac{y_1 \cot \theta + y_3 \tan \theta + x_3 - x_1}{\tan \theta + \cot \theta} \qquad (9.2)$$

$$tr_x = \frac{x_1 \tan\theta + x_4 \cot\theta + y_4 - y_1}{\tan\theta + \cot\theta},$$

$$tr_y = \frac{y_1 \cot\theta + y_4 \tan\theta + x_4 - x_1}{\tan\theta + \cot\theta} \qquad (9.3)$$

$$bl_x = \frac{x_2 \tan\theta + x_3 \cot\theta + y_3 - y_2}{\tan\theta + \cot\theta},$$

$$bl_y = \frac{y_2 \cot\theta + y_3 \tan\theta + x_3 - x_2}{\tan\theta + \cot\theta} \qquad (9.4)$$

$$br_x = \frac{x_2 \tan\theta + x_4 \cot\theta + y_4 - y_2}{\tan\theta + \cot\theta},$$

$$br_y = \frac{y_2 \cot\theta + y_4 \tan\theta + x_4 - x_2}{\tan\theta + \cot\theta} \qquad (9.5)$$

where θ is the object orientation.

Step 3 Calculate the pixel length L_p of the carrot (Fig. 9.5c).

Step 4 Build the relationships between the actual length L_a (cm) and the pixel length L_p by linear fitting (Fig. 9.5d):

$$L_a = 0.1115L_p - 0.3241 \qquad (9.6)$$

Normal carrots can be divided into four specifications: S, M, L, and 2L by the following:

$$grade = \begin{cases} S & L_a \leq 15 \\ M & 15 < L_a \leq 20 \\ L & 20 < L_a \leq 25 \\ 2L & L_a > 25 \end{cases} \qquad (9.7)$$

According to the grading requirements of carrot sales, normal carrots should hold a natural and uniform shape. As carrots are convex-shaped without a mutational edge in shape, a convex polygon approximation method was proposed to divide each specification into two levels (level 1 and level 2) to maintain consistency and uniformity in appearance. An

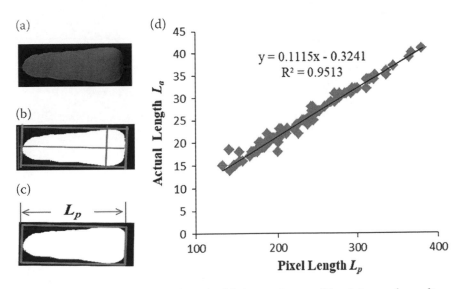

FIGURE 9.5 Fitting of carrot length. (a) Source image; (b) minimum bounding rectangle computation; (c) calculation of Lp; and (d) linear fitting of La and Lp.

external convex polygon is generated to approximate the contour of the carrots, and shape regularity R_s was defined as:

$$R_s = \frac{A_c}{A_p} \tag{9.8}$$

where A_c is the area of the carrot and A_p is the area of its external convex polygon. The larger R_s is, the more regular the carrot is. Besides, it is not affected by image size and carrot position, and has good adaptability.

A threshold θ can be set to classify carrots into two levels:

$$level = \begin{cases} 1 & R_s \geq \theta \\ 2 & R_s < \theta \end{cases} \tag{9.9}$$

9.2.5 Algorithm of defect detection and grading

The algorithm of defect detection and grading is described as below:

Algorithm 1 DEFECT DETECTION AND GRADING

Input: Source image *I*;
Output: Decision signal *s*;
    Initialization *s*= -1;
    Segment *I* into three images: left image *Ileft*, middle image *Imid* and right image *Iright*;
    for *i* in {*Ileft, Imid, Iright*} do
        Defect detection on image *i* by CDDNet: *dFlag*=CDDNet(*i*);
    if *dFlag*='defective'
        s = 0;
        exit for;
    end if
  end for
if(s = −1)
    Calculate the actual length *len* of image *Imid* by MBR fitting method;
    if *len*< = 15 then classify *Imid* into two levels (*s* = 1,2) by polygon Approximation method;
    else if *len*<=20 then classify *Imid* into two levels (*s* = 3,4);
    else if *len*< = 25 then classify *Imid* into two levels (*s* = 5,6);
    else then classify *Imid* into two levels (*s*=7,8);
    end
end

9.2.6 Evaluation standards

Five statistical parameters were calculated to evaluate the performance of the proposed model: accuracy, sensitivity, specificity, precision and F1-Score. They are specified by the following equations[21]:

$$Accuracy = \frac{TP + TN}{TP + TN + FP + FN} \tag{9.10}$$

$$Sensitivity = \frac{TP}{TP + FN} \tag{9.11}$$

$$Specificity = \frac{TN}{TN + FP} \tag{9.12}$$

$$Precision = \frac{TP}{TP + FP} \qquad (9.13)$$

$$F_1 - Score = \frac{2 * Precison * Recall}{Precision + Recall} \qquad (9.14)$$

where *TP*, *TN*, *FP*, and *FN* represent the number of true positives, true negatives, false positives, and false negatives, respectively.

9.3 RESULTS AND DISCUSSION

9.3.1 Effects of model parameters on CDDNet

9.3.1.1 Effect of batch size

The batch size represents the number of samples used in each step during the CNN training process. A suitable batch size can speed up the model training speed. To explore its effect on the accuracy and processing time of CDDNet, the proposed model was trained and validated on dataset 1 using different batch sizes (10, 20, 40, 80, and 160).

Fig. 9.6 presents the validation accuracy and training time with different batch sizes for 10 epochs. Validation accuracy is the evaluation of

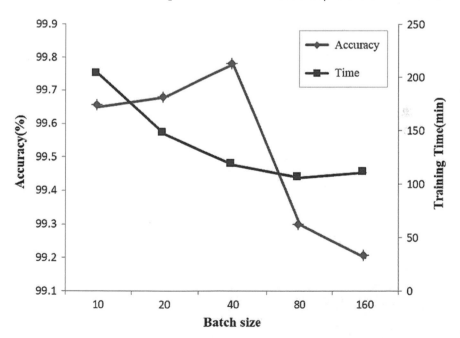

FIGURE 9.6   Validation accuracy and training time with different batch size.

model generalization ability[22]. The validation accuracy first increases and then decreases with the increase of the batch size. The highest accuracy is achieved when the batch size is 40. The training time decreases first and then has no significant change after batch size 40. For the trade-off between model generalization ability and training time, the value of batch size was set to 40.

9.3.1.2 Effect of the learning rate

The learning rate, as an important parameter in deep learning, determines whether and when the objective function converges to the local minimum. To select an appropriate value for our model, CDDNet was trained and validated on dataset 1 with different learning rates (0.05, 0.01, 0.005, 0.001, 0.0005, 0.0001, 0.00005, and 0.00001). Fig. 9.7 illustrates the validation accuracy of different learning rates. When the learning rate is too small (0.00001) or too large (0.05), the validation accuracy becomes relatively low. The best performance is achieved when the learning rate is 0.0005.

9.3.2 Performance of CDDNet to detect defective carrots

To evaluate the performance of CDDNet for carrot defect detection, we performed experiments on dataset 1 (binary classification) and dataset 2 (multi-class classification), and compared our method with some

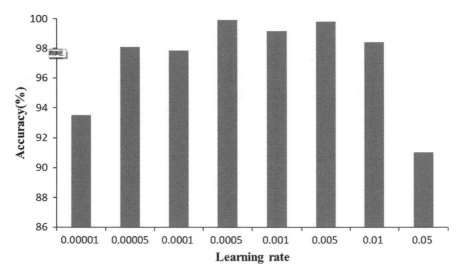

FIGURE 9.7 Validation accuracy of different learning rate.

prevalent methods, including the traditional handcrafted method[6], AlexNet[23], ResNet[24], and two lightweight models: MobileNet v2[25] and ShuffleNet[17].

The average values of the statistical parameters defined by Equations (9.10)–(9.14) were considered as the final performance of the model. All deep models were trained for 10 epochs by a cross-entropy loss function[26] and Adam optimizer with a learning rate of 0.0005. Experiments were performed on a computer (CPU: Intel(R) CoreTM i7-8700@ 3.20 GHZ, RAM: 8G; OS: Win10).

9.3.2.1 Performance evaluation of CDDNet for binary classification

Table 9.3 lists the average values of the statistical parameters (accuracy, precision, specificity, sensitivity and F1-Score), training time, processing time and model size for binary classification. Processing time represents the time spent on detecting an image.

The accuracy of the traditional method was only 93.12%, which was much lower than that of deep learning methods. Traditional carrot defect detection method[1,6] included several distinct steps, such as removing the background, detecting the region of interest[2] and designing feature for each kind of defect, which was complex and had poor flexibility. However, deep learning automatically extracted features of carrot defects, despite of little knowledge of what a defect was, the model worked quite well because it had an almost perfect sense of what a normal carrot was[27], which helped to reduce the error remarkably and achieved great advantages over traditional method.

The accuracy, precision, specificity, sensitivity and F1-Score of CDDNet were 99.82%, 99.90%, 99.81%, 99.64%, and 99.75%, respectively, which were only a slightly lower than those of ResNet (99.87%, 100%, 99.81%, 99.59%, and 99.80%). However, CDDNet had great advantages over ResNet in terms of training time, processing time and model size; its training time was only one-eighteenth that of ResNet, and its model size was only one-thirtieth that of ResNet. Compared to the original ShuffleNet, the model size of CDDNet was reduced; however, it still maintained good detection accuracy. Therefore, It is very necessary to select suitable model and appropriate network structure for carrot defect detection, which needs further study in the future. In general, CDDNet achieved outstanding performance both in detection accuracy and model complexity.

TABLE 9.3 Experimental results of the defect detection performance of binary classification

Methods	Accuracy (%)	Recall (%)	Specificity (%)	Precision (%)	F$_1$-score (%)	Training time(min)	Processing time(ms)	Model size(MB)
Handcrafted	93.12	–	–	–	–	–	108.4	–
AlexNet	98.86±0.60	98.98±0.76	98.80±0.97	97.52±1.98	98.23±0.91	254.2	54.9	214.1
ResNet50	**99.87**±0.09	**100.00**	**99.81**±0.11	99.59±0.32	**99.80**+0.13	2044.2	85.2	86.5
MobileNet v2	98.94±0.19	**100.00**	98.44±0.29	**96.76**±0.59	98.35±0.30	256.7	49.2	8.4
ShuffleNet	99.77±0.27	99.96±0.06	99.68±0.32	99.33±0.57	99.64±0.33	125.0	25.5	3.3
CDDNet	99.82±0.11	99.90±0.10	**99.81**±0.19	99.64±0.32	99.75±0.17	**118.5**	**26.9**	**2.7**

Time efficiency is an important aspect to be considered in online carrot grading. Our previous work[6] could detect about 12 carrots per second. It can be seen from Table 9.3 that CDDNet could detect approximately 40 carrot images per second, which make it feasible to meet the demand of real-time detection of carrot defects.

9.3.2.2 Performance evaluation of CDDNet for multi-class classification

Table 9.4 illustrates the comparative results among our method and some prevalent methods for multiclass classification. In this task, CDDNet achieved an overall accuracy of 93.01%. Although ResNet model achieved the highest overall accuracy (94.28%), the training time was much longer. Regarding the trade-off between time and accuracy, our model demonstrated an overall advantage over other models on the multiclass classification task.

There was a large difference in the accuracy among different kind of defects. Generally, the accuracy for normal and abnormal carrots was high; while the accuracy for fibrous root and bad spot was relatively low, especially bad spot. There may be two reasons:

1. Sample imbalance problem. There was a large difference in sample size among different categories; there were 1385 abnormal samples; however, there were only 294 fibrous-root samples. The sample imbalance caused the classification model to have a serious bias. Some strategies can be used to solve the sample imbalance problem, such as expanding data set[28], oversampling[29] and undersampling[30], and generating data samples manually[28]. In the future, we will enlarge image samples by collecting more carrot samples or using data augmentation technique[31] to further improve the performance of the detection model.

2. Diversity of defect appearance. There were several factors leading to bad spots, such as mechanical damage, decay, bruises and attachments, resulting in a large difference in appearance. Due to the large difference in phenotype, it is difficult for the CDDNet to extract effective features, which might be another reason for the low detection accuracy of bad spot. One possible solution is to identify each of these possible situations as a separate category.

TABLE 9.4 Experimental results of the multiclass detection performance

Methods	Accuracy (%)					Training time (min)
	Normal	Abnormity	Fibrous root	Bad spot	Total	
Handcrafted	91.21	93.93	95.52	81.18	91.01	–
AlexNet	95.80±3.62	90.40±9.80	93.26±3.41	84.34±11.14	91.28±4.10	56
ResNet	97.12±2.16	95.38±1.76	90.68±8.42	88.52±3.10	**94.28±0.60**	447
MobileNet v2	98.18±0.88	91.68±2.93	90.90±8.40	80.98±6.94	91.65±1.70	93
ShuffleNet	98.08±0.74	92.08±4.80	95.75±2.98	83.13±7.85	92.63±1.46	44
CDDNet	97.54±0.95	97.02±2.36	91.14±7.22	76.96±6.97	93.01±1.60	**43**

9.3.2.3 Comparison with previous studies

Comparing the results of CDDNet with those of other research is a challenging work due to lack of public datasets as benchmark in this field. Published research on defect detection for agricultural products[1,2,12,13] usually use their own datasets, which makes it difficult to compare our results with those of related work. In the previous work, Zhu et al.[14] transferred pre-trained AlexNet to identify the carrot appearance quality and used the same dataset as ours. However, their model parameters are different from ours. To make a fair comparison, for all deep learning models, we used the same model parameters for training all deep learning models, and performed the experiment using pre-trained AlexNet as Zhu et al. instead of comparing with their experimental results directly. So in this paper, we compared the proposed method with our previous work and some prevalent deep learning methods: AlexNet, ResNet and ShuffleNet on our own dataset. Although without a proper comparison, considering the limited research studies using computer vision for carrot defect detection[1,5], most research achieve an overall accuracy below 91% on their datasets.

The proposed CDDNet performed defect detection well without manually defined features or any other preprocessing. Therefore, the model should achieve similar performance level when applied to other agricultural products if sufficient dataset is available.

9.3.3 Evaluation of carrot grading method

9.3.3.1 Evaluation of the MBR fitting method

In this experiment, 918 carrots covering four specifications (S, M, L, 2L) were selected as experimental samples. First, these carrots were graded into four specifications (S, M, L and 2L) by 4 skilled workers as they usually do. Then, images of these carrots were acquired by the computer vision system and graded by the MBR fitting method. Grading results were validated by measuring the actual carrot length with a ruler, as presented in Fig. 9.8.

Experimental results showed that our grading method achieved an overall accuracy of 92.8%, which was much higher than that of the manual method (83.1%). The grading accuracy of the manual method varied greatly among different specifications, and the accuracy of S and M was only 78.4% and 69.9%, while accuracy of grades L and 2L was both above 90% (90.7% and 95.8%). Our method achieved accuracy of

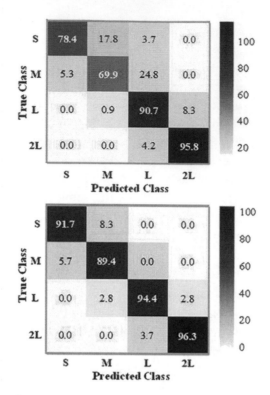

FIGURE 9.8 Confusion matrixes of (a) manual method and (b) the proposed method.

91.7%, 89.4%, 94.4%, and 96.3%, respectively, for S, M, L, and 2L, which demonstrated more reliability and accurate grading performance than manual method.

As there is no related work on carrot grading by size fitting, we only compared our method with manual method in this paper. It is obvious that manual method has such shortcomings as low accuracy and instability. Our method had obvious advantages not only in grading accuracy but also in reliability, which greatly improved the accuracy and efficiency of carrot grading.

9.3.3.2 Evaluation of the convex polygon approximation method
To evaluate the performance of convex polygon approximation method, we performed experiment on dataset 4 (Table 9.5). It can be seen that convex polygon approximation method achieved an overall grading

TABLE 9.5   Results of convex polygon fitting method

Items	Correct	Error	Total	Accuracy (%)
Level 1	652	31	683	95.4
Level 2	221	14	235	94.0
Total	873	45	918	95.1

accuracy of 95.1% and showed a good ability to measure the degree of shape regularity.

Current related studies[1,5,6,14,15] on carrots mainly focused on surface defect detection, and there are few research reports on carrot grading based on computer vision. Although our method was simple, it provided an effective method for carrot grading. There are some criteria in grades and specifications of the carrot, such as shape uniformity, surface smoothness, colour uniformity and mature degree. However, most of these criteria are qualitative and not easy to quantify. In the future, we will further study the measurement of carrot grading criteria, and take more factors into consideration, such as surface smoothness, colour uniformity, and mature degree.

9.3.4 Practicability of the proposed approach

In this chapter, according to the actual quality processing and grades and specifications of the carrot, we proposed defect detection and carrot grading methods. Defective carrots (abnormity, cracks, fibrous roots, fractures, and bad spots) were first sorted out by binary classification of CDDNet, and then normal carrots were classified into different specifications and grades. Compared with the previous work, our system was more consistent with the actual process and realized full automation of carrot quality inspection and grading, which could be easily applied to the actual conditions.

The CDDNet could detect most kinds of defects affecting the appearance quality of the carrot, including abnormity, cracks, fibrous roots, fractures, and bad spots caused by damage, erosion and impurity, and achieved a detection accuracy of 99.82% for binary classification and could detect about 40 carrots per second. The grading method classified the carrots into 4 specifications and graded each specification into 2 levels with grading rates of 92.8% and 95.1%. So the proposed method could meet the requirement of practical application both on accuracy and efficiency.

9.4 CONCLUSIONS

In this chapter, we proposed carrot quality inspection methods and grading methods based on computer vision and deep learning. A lightweight network (CDDNet) is developed for on-line carrot defect detection based on ShuffleNet and transfer learning, and the carrot grading method is proposed based on MBR fitting and convex polygon approximation. Experimental results demonstrated that the proposed CDDNet achieved a detection accuracy of 99.82% for binary classification and 93.01% for multi-class classification, and the grading accuracy of MBR fitting and convex polygon approximation was 92.8% and 95.1%, respectively, which could meet the requirements of actual production. Our research has proven the feasibility of deep learning for carrot defect detection, and provided a practical way for online defect detection and grading of carrots.

REFERENCES

1. Xie, W., Wang, F., and Yang, D. 2019. Research on carrot surface defect detection methods based on machine vision. *IFAC-Papers OnLine* 52(30): 24–29.
2. Nasiri, A., Omid, M., and Taheri-Garavand, A. 2020. An automatic sorting system for unwashed eggs using deep learning. *Journal of Food Engineering* 283(1):110036.
3. Nturambirwe, J. f. i, and Opara, U. L. 2020. Machine learning applications to non-destructive defect detection in horticultural products. *Biosystems Engineering* 189:60–83.
4. Azdc, A. HeHf, A., and Jaf, B. 2020. Computer vision based detection of external defects on tomatoes using deep learning. *Biosystems Engineering* 190:131–144.
5. Han, Z., Deng, L., Xu, Y., Feng, Y., Geng, Q., and Xiong, K. 2013. Image processing method for detection of carrot green-shoulder, fibrous roots and surface cracks. *Nongye Gongcheng Xuebao/Transactions of the Chinese Society of Agricultural Engineering* 29(9):156–161.
6. Deng, L., Du, H., and Han, Z. 2017. A carrot sorting system using machine vision technique. *Applied Engineering in Agriculture* 33(2): 149–156.
7. Du, H., Deng, L., Xiong, K., and Han, Z. 2015. Quality grading system and equipment of carrots based on computer vision. *Journal of Agricultural Mechanization Research* 1:196–200.
8. Lecun, Y., Bengio, Y., and Hinton, G. 2015. Deep learning. *Nature* 521(7553): 436.

9. Jiang, B., et al. 2019. Fusion of machine vision technology and AlexNet-CNNs deep learning network for the detection of postharvest apple pesticide residues. *Artificial Intelligence in Agriculture* 1:1–8.

10. Sun, Y., Lu, R., Lu, Y., Tu, K., and Pan, L. 2019. Detection of early decay in peaches by structured-illumination reflectance imaging. *Postharvest Biology and Technology* 151:68–78.

11. Liu, Z., Yong, H., Cen, H., and Lu, R. 2018. Deep feature representation with stacked sparse auto-encoder and convolutional neural network for hyperspectral imaging-based detection of cucumber defects. *Transactions of the Asabe* 61(2):425–436.

12. Thuyet, D. Q., Kobayashi, Y., and Matsuo, M. 2020. A robot system equipped with deep convolutional neural network for autonomous grading and sorting of root-trimmed garlics. *Computers and Electronics in Agriculture* 178:105727.

13. Fan, S., Li, J., Zhang, Y., Tian, X., and Huang, W. 2020. On line detection of defective apples using computer vision system combined with deep learning methods. *Journal of Food Engineering* 286:110102.

14. Zhu, H., Deng, L., Wang, D., Gao, J., and Han, Z. 2019. Identifying carrot appearance quality by transfer learning. *Journal of Food Process Engineering* 42(10): e13187.

15. Zhu, H., Yang, L., Sun, Y., and Han, Z. 2021. Identifying carrot appearance quality by an improved dense CapNet. *Journal of Food Process Engineering* 44(1):e13586.

16. Deng, L., Li, J., and Han, Z. 2021. Online defect detection and automatic grading of carrots using computer vision combined with deep learning methods. *LWT- Food Science and Technology* (2): 111832.

17. Zhang, X., Zhou, X., Lin, M., and Sun, J. 2018. Shufflenet: an extremely efficient convolutional neural network for mobile devices. *In Proceedings of the IEEE conference on computer vision and pattern recognition* (pp. 6848–6856).

18. Weiss, K., Khoshgoftaar, T. M., and Wang, D. 2016. A survey of transfer learning. *Journal of Big Data* 3(1):1–40.

19. Kala, J. R., Viriri, S., and Tapamo, J. R. 2014, October. An approximation based algorithm for minimum bounding rectangle computation. *In 2014 IEEE 6th International Conference on Adaptive Science & Technology (ICAST)* (pp. 1–6).

20. Tomé, A. R., and Miranda, P. M. A. 2004. Piecewise linear fitting and trend changing points of climate parameters. *Geophysical Research Letters* 31(2): L02207. 1–L02207. 4.

21. Fawcett, T. 2006. An introduction to ROC analysis. *Pattern Recognition Letters* 27(8):861–874.

22. Lorbert, A., and Ramadge, P. 2010, March. Descent methods for tuning parameter refinement. *In Proceedings of the Thirteenth International Conference on Artificial Intelligence and Statistics* (pp. 469–476).

23. Krizhevsky, A., Sutskever, I., and Hinton, G. E. 2012. ImageNet classification with deep convolutional neural networks. *Advances in Neural Information Processing Systems* 25:1097–1105.

24. He, K., Zhang, X., Ren, S., and Sun, J. 2016. Deep residual learning for image recognition. *Proceedings of the IEEE Computer Society Conference on Computer Vision and Pattern Recognition*: 770–778.

25. Sandler, M., Howard, A., Zhu, M., Zhmoginov, A., and Chen, L. C. 2018. MobileNetV2: Inverted residuals and linear bottlenecks. *Proceedings of the IEEE Computer Society Conference on Computer Vision and Pattern Recognition*: 4510–4520.

26. Drozdzal, M., Vorontsov, E., Chartrand, G., Kadoury, S., and Pal, C. 2016. The importance of skip connections in biomedical image segmentation. *In Deep learning and data labeling for medical applications* (pp. 179–187). Springer, Cham.

27. Azdc, A., HeHf, A., and Jaf, B. 2020. Computer vision based detection of external defects on tomatoes using deep learning. *Biosystems Engineering* 190:131–144.

28. Chawla, N. V. 2009. Data mining for imbalanced datasets: an overview. *Data mining and knowledge discovery handbook*: 875–886. Boston: Springer.

29. Zhang, X., Wang, Z., Liu, D., and Ling, Q. 2019, May. Dada: Deep adversarial data augmentation for extremely low data regime classification. *In ICASSP 2019-2019 IEEE International Conference on Acoustics, Speech and Signal Processing (ICASSP)* (pp. 2807–2811).

30. Hensman, P., and Masko, D. 2015. The impact of imbalanced training data for convolutional neural networks. *Degree Project in Computer Science*, KTH Royal Institute of Technology.

31. Shorten, C., and Khoshgoftaar, T. M. 2019. A survey on image data augmentation for deep learning. *Journal of Big Data* 6(1):1–48.

Conclusions

SUMMARY

Aiming at the central issues of how to represent, store, and extract visual information in the human brain, inspired by the mechanisms of biological visual perception and human memory, this book discusses and studies the methods of visual information representation and memory modelling based on the related research results of cognitive neuroscience. The main works of this book are as follows:

1. Based on the research findings of biological visual perception mechanism and visual cognitive modelling, a five-layer biological visual perception model sparse HMAX was established by combining multi-firing K-means, non-negative sparse coding, and the HMAX model to better simulate the biological characteristics of visual neural cells in V2 and V4 regions of human visual pathway. The multi-firing K-means was used to simulate the characteristics of V2 neurons, and NNSC was introduced into S2 layer to better explain the response characteristics of V4 neurons.

2. Considering the fact that the HMAX model is more sensitive to rotation, the characteristics of low-level retina and external geniculate neurons are not considered, and the biological characteristics of high-level neurons are not fully described, according to the characteristics of neurons in various regions of the cerebral visual cortex, an invariant feature extraction model is established based on biological visual perception combined with the GLoP filter

DOI: 10.1201/9781003281641-10

and multi-manifold sparse coding. Contrast regularization is used to process visual input, and the GLoP filter is used to simulate simple cells in V1 area. Considering the sparse excitation characteristics of V4 neurons and manifold visual perception methods, multi-manifold sparse coding is used to simulate the response of nerve cells in the cerebral cortex of V4 area. In addition, a template learning method based on multi-manifold dictionary learning is proposed.

3. In order to realize rapid storage and extraction of visual information, based on the relevant research results of cognitive neurology, Increment Pattern Association Memory Model (IPAMM) is proposed, and the Leabra learning mechanism is introduced into the pattern-associative memory model to replace the simple Hebb learning rules and assign separate weights to different categories to achieve increment learning and avoid interference to a certain extent. Experimental results show that this method can complete the image classification task, especially for small training sets, and it has better recognition performance, and has higher time efficiency, and can better meet the requirements of practical applications.

4. Friston's free energy theory provides a powerful theoretical framework to describe how the brain perceives and remembers external stimuli, but there are few specific application results. Based on this, this book combines Friston's free energy theory with RBM to realize the learning and memory of visual information.

5. In order to detect and identify crop pests quickly, the proposed visual perception model and visual attention mechanism are combined for the detection and identification of crop pests. First, detect pests and extract regions of interest based on the natural statistical significance model. Then, extract invariant features based on the visual perception model and LCP algorithm. Finally, input the extracted features into the support vector machine for pest identification.

6. In order to realize automatic quality inspection and grading of carrots, a lightweight deep learning model (CDDNet) was constructed to detect surface defects based on ShuffleNet and transfer learning. Also, carrot grading methods were proposed based on minimum bounding rectangle (MBR) fitting and convex polygon approximation.

FUTURE WORK

1. The biological visual perception model proposed in this book can extract the invariant features of the image and has achieved good performance in image classification tasks, but its recognition accuracy still has a large gap with the current popular deep network. How to learn from related theories and methods of deep learning to further improve model performance, will be the main work of the next step.

2. The IPAMM model can achieve better recognition performance when there are fewer training samples. However, when the number of training samples is large, the model has no obvious advantages, so it is necessary to consider how to improve recognition accuracy in the case of more training samples. At the same time, the IPAMM model only contains one learning layer. Although a simple structure can achieve better time efficiency, adding additional hidden layers may further improve the classification performance of the model.

3. The memory model based on free energy theory and RBM proposed in this book uses RBM as a probability generation model to simulate Friston's brain cognition and inference procedure. However, this model is only used for image classification and does not describe the image storage and extraction process in detail. How to realize the free energy theoretical framework of brain perception and prediction needs to be further explored in theory and practice.

4. While studying new models and algorithms of visual information cognition, more attention should be paid to practical applications. In the future, we will apply these methods to more fields and solve more practical problems.